TROPICAL FOLIAGE PLANTS

A Grower's Guide

ABOUT THE AUTHOR:

Lynn Griffith is President of A & L Southern Agricultural Laboratories, Inc., an agricultural testing and consulting firm located in Pompano Beach, Florida. He and his co-workers provide laboratory testing and consultation for thousands of clients throughout Florida, the Caribbean, and Latin America. He also provides horticultural consulting services to clients in 14 countries and conducts contract research for a number of fertilizer, chemical, and potting soil companies.

The author has written numerous technical articles for trade journals over the last 15 years. Educated at Duke University, he lives in West Palm Beach, Florida, where his hobbies include gardening, music, and Cajun cooking. His e-mail address is lgriff6250@aol.com.

TROPICAL FOLIAGE PLANTS

PLANTS

A Grower's Guide

BY

LYNN P. GRIFFITH, JR.

Ball Publishing

Batavia, Illinois USA

Ball Publishing
335 North River Street
Batavia, Illinois 60510 USA

© 1998 by Ball Publishing. All rights reserved.
Printed in the United States of America
03 02 01 00 99 98 5 4 3 2 1

Library of Congress Cataloging-in-Publication Data

Griffith, Lynn P., 1956–
    Tropical foliage plants : a grower's guide / by Lynn P. Griffith, Jr.
        p.   cm.
    Includes bibliographical references and index.
    ISBN 1-883052-16-5
    1. Foliage plants. 2. Tropical plants. 3. Greenhouse plants.
I. Title.
SB431.G75   1997   97-31847
635.9'75—dc21   CIP

DEDICATION

To Lynn Porter Griffith, III,

"Little One"

Contents

viii Contents

FOREWORD

The Early Years of Foliage in the United States

BY JIM VOSTERS, SR.

1926-1945

Before World War II, foliage production took place in glass greenhouses. Greenhouse growers in most major cities had small foliage sections. Their primary interest, however, was in an assortment of flowering plants for major holidays--foliage was a sideline. Foliage in the early 1900s was mainly *Ficus elastica* (the famous rubber tree), ferns, and *Sansevieria*. Aglaonemas were unknown except for special foliage growers in Belgium, the Netherlands, Germany, and France. In the United States the major foliage producers were the Robert Craig Co. of Philadelphia, Hans Plath Co. of San Francisco, Joseph Heacock Co. of Philadelphia, and the Oechslin Co. of Chicago.

    Sansevieria laurentii was brought to the United States by the Macaw Brothers of Norwood, Pennsylvania. They made a lot of money when they planted 15 acres in Cutler Ridge, Florida, near Miami. This was in 1926, and land was $20 an acre, labor $1 a day. They got a dollar a leaf wholesale, FOB Cutler Ridge, Florida, for *Sansevieria*. Incidentally, there was no federal income tax on earnings from agriculture until 1943!

1946-1951

During World War II most glasshouses had been destroyed in Germany and the Netherlands. The northwestern part of Belgium in the Ghent area escaped serious damage, and that was the location of Flandria and the Van Houte Co., both large foliage producers. The big producer in England was the Thomas Rochford Company. Alas, all of these are now gone. In the United States no foliage specialists are 100 years old, although there are growers who began as foliage specialists and switched crops as the market changed through the years.

x Tropical Foliage Plants: A Grower's Guide

In 1946 the market, as before, was dependant on retail flower shops, whose real interests were cut flowers, ceramics, and dried flowers. Plants were a sideline. Most growers were reluctant to specialize in foliage because the market was so limited. However, from 1946 to 1949, many new homes were built for families of returning soldiers and this aided the market for small foliage plants for the home.

By 1948 I decided that my future lay in the direction of foliage plants. Two things made me decide this. The first was that the return per square foot for 2-1/4-inch, 3-inch, and 4-inch pots was far greater than for large flowering plants, such as hydrangeas and azaleas. The second thing that directed my career was the F. W. Woolworth Co. The buyer for its 3,200 stores was wise in seeing that they could sell thousands of small potted foliage plants without much shrinkage. Woolworth was in control of the foliage market until 1960.

Despite our best efforts to sell to its competitor, the S. S. Kresge Co., with about 1,000 stores, we were not very successful. The Kresge buyer wanted consigned sales; that is, if a plant didn't sell, the producer gave Kresge its money back. I refused to be a part of such a marketing agreement, because the clerks would not have any incentive to take the slightest care of the plants. In major cities like Philadelphia, Boston, and New York, truck delivery from grower to store counter was required. This was true all over the United States.

1952-1960

In 1952 the Florida growers began to ship to store door by Railway Express, whose service was terrific in the 1950s. Shipments from Miami or Apopka to destinations like Kansas City took at most three days, which was quite acceptable for the hottest items at the time, which were *Philodendron cordatum, Pothos wilcoxii,* and Boston fern.

Obviously, the cost of production in Florida was lower than in northern glasshouses. Labor was cheaper in the South, and the capital investment was much smaller. In 1952 the Woolworth buyer urged northern suppliers to move to Florida to boost production. Strangely, only a handful of foliage growers in the Northeast and Midwest took the plunge. I was one of them, along with Lex Ritter of New Jersey, Lou Super of Brooklyn, New York, John Good of Springfield, Ohio, and later Larry Heinl of Toledo, Ohio. All of us, except for Super, were Woolworth suppliers, and the race was on.

The Florida foliage market from 1946 to 1952, when the growers mentioned arrived from the North, was very different from today. Apopka was the Central Florida producer for the variety stores. At that time supermarkets and discount houses would not touch live plants or flowers for any reason. Producers such as Harry Ustler and John Massik would grow Boston ferns, wrap the roots in wax paper and hold the wrapping with a rubber band. In 1950 literally boxcar loads of these ferns were shipped by Railway Express. Five-and-dime stores considered the ferns a specialty. In the Miami area from 1946 to 1952, there were four major shippers: Dan Greer of D.J. Greer Nursery, Hialeah, (who was the biggest), along with Paul Oskierko of Oskierko Nursery, Homestead, Oscar Nelson of Opa Locka, and Hugh Layler of Caribbean Nursery, Opa Locka. Greer concentrated on variety stores. The others sold their products to owners and operators, who resold to local stores in their areas. Winter Garden Ornamental Nursery in Winter Garden, Florida, was also a very large supplier.

Greenhouse buildings were made of wooden slats on a wooden frame. Heat was supplied by orange grove heaters. If it rained, the crop could be ruined. Dan Greer bought land in Hialeah in the industrial section by the railroad tracks. He made slathouse frameworks out of railroad ties that he got free from the railroad and made the slats out of used orange crates. With these nearly free greenhouses, Greer had tremendous success and controlled the Woolworth business in the 1950s. He later sold the land for industrial warehouses.

Clay pots were the favorite containers for plants, but the advent of plastic pots changed the entire business for the grower. Freight to the Woolworth store in Boston, for instance, was payable by the store. They wanted the product and were willing to pay the freight. Railway Express charged by weight and distance. There was no commercial trucking of foliage in those days, so because of weight plastic pots were the only way to go. In May 1952 newcomers from the North joined the Florida Native Producers, which gradually evolved into the trade organizations that exist today.

In the late 1950s two new products became available and changed the industry. One was saran cloth in wide widths. So little was known about saran shade that salesmen told the growers to install it as tightly as possible. What they didn't know was that saran shrank 8% in the sun, and after exposure it would pull the end post right out of the ground. Normal post spacing was on 10-foot centers, but with an ever-increasing effort to build as cheaply as possible, the post spacing soon became 40 feet, and the first major wind storm blew many buildings to pieces. The other great change was sheet polyethylene in wide widths, which could shield the production facility from the elements. Polyethylene was probably the single most important thing that could have developed, as it became a cheap means of maintaining greenhouse temperatures and keeping out the rain.

Another defining moment was the advent of trucking companies for foliage. Johnny Brown, the founder of this business, initially had one truck in Minneapolis that would haul hay to the Florida racetracks. Alex Laurie, a retired dean of horticulture at Ohio State University, had started a foliage business in Eustis, Florida. He met Johnny Brown, who was desperate to get a back haul, and so began the plant-hauling industry. The truckers, however, could not deliver just two boxes to a five-and-dime's door. Truck deliveries had to be made to supermarket produce warehouses, and then daily deliveries of the produce would also deliver plants to the stores.

1961-1970

Railway Express closed down in 1965, and by that time, after much effort and cajoling, a few more supermarkets were beginning to handle plants. They would not touch cut flowers, because they were frightened to death of cut flowers. We growers knew that we absolutely had to have supermarkets or we would choke on our own production. Remember that in 1960 Home Depot, Wal-Mart, and Kmart were not yet available as potential plant outlets. Furthermore, discount stores did not individually warehouse, and they needed store delivery.

While the Florida growers were expanding, a sizeable foliage producing area began in southern California. The undisputed leaders were Bob Weidner of Buena Park Greenhouses in Lattabra, Schneirow Nursery in San Diego, and Nurserymen's Exchange in Half Moon Bay. Weidner was an innovator and a stickler for quality. In Brownsville, Texas, there were two large producers, the Rio Grande Nursery and Neal Robinson Greenhouses. The California product went mostly to California, Washington, and Oregon. The Texas production went to the Midwest, and the Florida products mostly to the East Coast. Florida was by far the leader in both volume and dollar value. In 1960, after Cuba's Castro revolution, the Costa family emigrated to Homestead, Florida, and became a very major player in the Florida foliage scene.

These were the days when offshore nurseries began in Guatemala, Jamaica, Costa Rica, Honduras, and the Dominican Republic. Since before WWII, Puerto Rico had one very large producer, Charlie Pennock of North South Nursery. He shipped primarily *Philodendron cordatum* and *Dracaena* canes. Robert Craig Co. had a nursery in Santurce, as well, and that is where *Dracaena* Janet Craig began. Because of Quarantine-37, plants in media were not allowed entrance into the United States. This meant that offshore nurseries, with their incredibly low costs, could not ship finished products, only bare-root cuttings and air layers. If Quarantine-37 had not kept finished foliage from foreign producers out of the country, the foliage industry could have gone the way of the U.S. cut flower industry.

Many people became interested in new foliage varieties. Sometimes people, such as Bob Wilson of Fantastic Gardens in Miami, were interested in all foliage plant types. Some, like Nat Deleon of Miami, had a special interest in a few plants, such as aglaonemas or bromeliads. New varieties came to these people by mutation, breeding, and collecting. Others traveled the world from their glasshouses in a temperate climate to the rain forests of Ceylon, Indonesia, and other tropical areas. Alfred Graf, the author of *Exotica*, was the most experienced traveler of all. All of their efforts led to a greater diversity of houseplants on the market.

1971-1980, THE GOLDEN YEARS OF FOLIAGE PLANTS

By 1974 there was a foliage plant boom. The boom was so huge that growers couldn't believe what they were seeing. It was impossible to fill the demand for some items. This attracted the attention of large players, such as the United Fruit Company, Pillsbury, and Arvida, who thought they could make an easy killing. Arvida, the brainchild of Arthur Vining Davis, was the first corporate player in foliage. In 1960 Arvida had a large nursery in the location of what is now Dadeland Shopping Mall in Miami. Many large, publicly held companies outside of the industry wanted to cash in on the boom, but not one has stayed with it to this day. The producing side does not seem to work for conglomerates.

Brokers of plants, seeds, and flowers, like Ball Seed Co., Fred C. Gloeckner & Co., Skidelsky, and McHutchinson & Co., played an important part in distributing foliage plants. They provided advice and even capital to some growers to get a supply of plants on a steady basis. The salesmen from these companies covered every village and town in the United States. No grower was big enough to have that in-house sales coverage.

A few more northern growers, notably Herman Engelmann, moved south to Florida and created a very fine, large producing facility. Every grower wanted to find a niche for a particular specialty in the market. Usually, unless the product was patented, the exclusivity didn't last very long. Imitation is the sincerest form of flattery. Many new developments were immediately copied, and each copier tried to underprice the other. In later years, though, many foliage growers, unable to price their products properly and make a proper profit, were having difficulty staying in business.

1981-1986

This was the age of growing techniques that changed the industry forever. Innovations included systems for controlling the environment, soil testing services, automatic rolling benches, computers, and, most importantly, the beginning of tissue culture. Ray Oglesby of Hollywood, Florida, was the first commercial tissue-culture producer of foliage in the United States. He put everything he had into this concept, at great risk, and was successful.

TODAY

Business methods are so different now. Computers are used everywhere. Business is transacted by fax and e-mail instead of by phone and letters. Production is now highly automated, but are the sales keeping up with production? More effort is needed to market the product. We've come a long way in 50 years. I invite the new people entering our business to buy land for their production facilities not for a remote location at a very cheap price, but rather for a location in the path of city growth. The last big profit for a foliage grower comes from selling the land to a developer. In the meantime, find your niche, grow quality, and market aggressively.

AUTHOR'S NOTE:

Jim Vosters, Sr., died on February 3, 1997. He was born July 25, 1922, in Secane, Pennsylvania, where he started his first nursery in 1946. He moved to Miami, began Vosters Nurseries Inc. in 1952, and remained there until 1989. Jim served as president of the Society of American Florists from 1973 to 1975 and was inducted into the SAF Hall of Fame in 1986.

PREFACE

When I started growing foliage plants in 1976, there was very little information available on how to actually grow them. Most of us learned by trial and error, making many mistakes along the way. When I found people who seemed to know what they were talking about, I pumped them with endless questions and remembered what they said. I discovered a number of differences in the way growers do things, and there were and still are a number of misconceptions and differences of opinion about growing foliage.

I therefore decided to get into the information side of the business and have been fortunate to provide consultation for thousands of growers in many parts of the world. I have learned a lot seeing different types of plants grown in different environments and with dissimilar production techniques. Soils, fertilizers, varieties, and philosophies differ wherever you go, yet at the same time, there are fundamental rules which apply in virtually all situations. In fact, at the end of this book I will reveal to you, dear reader, THE SECRET to growing foliage plants successfully.

I thought about writing this book for 10 years before finally deciding to do it. Many clients have asked me, "Is there a book out there with some of this good technical information in it?" I would always have to tell them that there is quite a bit of good research and technical information on producing foliage, but much of it is scattered about in obscure journals that growers don't have time to read, and the articles may be difficult to understand, anyway. There are a lot of plant books out there, but few of them are really good. Too many of them are written by people who have never grown a crop in their lives.

We owe a great debt of gratitude to the researchers who have made discoveries and helped us forge ahead in the foliage industry. At the same time, there is a great body of practical knowledge that the growers themselves have accumulated. I have picked up useful bits of information from brilliant Ph.D. research scientists as well as from farm laborers in the middle of a jungle who have never seen the inside of a school. Growing is an art as well as a science, and neither can survive without the other. I have tried to take this diverse information and assemble it into a format that will be a step forward in understanding foliage plants and growing better ones.

ACKNOWLEDGMENTS

I would first of all like to thank Joan Donlon for patiently typing and retyping every word of this wretched manuscript and Becky Van Acker for helping to organize it. A hearty muchas gracias to Ed Clay, Marshall Horsman, Bill Lewis and John Gatti for photographs and manuscript reviews. I am indebted to Jim Vosters for his effort in writing an enlightening foreword. I found his personal experiences and insight to be invaluable. I greatly appreciate all those who submitted photographs for the book and the great people at Ball Publishing for their ideas, cooperation, and encouragement.

I would like to thank all my clients and everyone from whom I have ever learned about plants, as well as Hugh Poole and Don Ankerman for giving me a chance. Special thanks to Albert King, whose records provided much-needed intervals of stress relief. Thanks to Suzie, for everything.

And finally, a special word of gratitude to you, the reader, for supporting the effort. I hope you find it well worth it.

INTRODUCTION:

HOW TO USE THIS BOOK

A plant book will not make you a great grower, in the same way that a cook-book will not make you a great chef. However, if you desire to be more efficient in growing or maintaining plants, I have attempted to create a handbook that will supply you with much of the necessary information. This book is not intended to be read from cover to cover; since each section focuses on one or more of the major varieties in the foliage trade, I suspect that readers will want to jump around from one foliage variety to another foliage variety based on their immediate needs.

We should briefly define the term *foliage plant* as used herein. I am considering any plant intended to be placed in the home or office for an extended period to be a foliage plant. Certainly, some foliage plants flower, and many of the varieties in this book have important uses other than as foliage plants. I have not tried to write about *every* plant grown for foliage, just the more important ones.

Each foliage variety is organized into the following sections: *habitat*, *uses*, *varieties*, *propagation*, *culture*, *nutrition*, *diseases*, *insect and mite pests*, *disorders*, *tricks*, and *interior care*. An extensive color plate section illustrates numerous disorders. Scattered throughout the book are "pearls of wisdom," which are general tips and information useful to foliage growers and interior maintenance technicians.

HABITAT

Most books tell you what country a plant comes from originally, but that alone is really not useful information. You will find that many of the keys to understanding how a plant responds are related directly to how it functions in its native environment. I have indicated in the book not only the country of origin but what sorts of conditions the foliage plant is accustomed and adapted to. I contacted growers and horticulturists in many countries to obtain the most specific information possible. If you are serious about growing a foliage plant as your profession or even just as a hobby, I strongly urge you to visit its native habitat. You will be a better grower having done so.

Uses

Foliage plants range in size from tiny to tremendous. This section outlines the major pot sizes and the more common types of production. Information on how these plants are used in interiorscapes is provided. Landscape use references are also given because many growers in tropical areas end up selling some of their production to the landscape industry.

Varieties

Where applicable I have tried to discuss the most commonly grown varieties of each foliage plant. Granted, that may vary somewhat from country to country. New varieties are entering the marketplace all the time. One may stay in the trade for a century, whereas another is a flash in the pan. I have tried to avoid talking about foliage varieties that have yet to prove themselves. I encourage growers to try some of the newer varieties coming out, as breeding efforts are improving all the time and variety mixes are constantly evolving.

Propagation

Some foliage growers do their own propagating, while others purchase plugs or liners of various types. The various methods of propagation are outlined, including the commercially viable methods for the professional grower as well as smaller scale methods for the serious hobbyist. I have tried to be as specific as possible, especially concerning the commercial production techniques. Growers need to know exactly how and when to plant a seed or root a cutting, and what to expect.

Culture

This section discusses such factors as light levels, environmental conditions, watering practices, and potting mixes. Again, I have tried to be as specific as possible, but it is difficult to recommend a precise potting mixture, for example, without knowing pot size, irrigation method, and type of greenhouse. You will find in virtually all types of horticulture that more than one method works. I have tried to remain sensitive to the fact that the best way to grow aglaonemas in a Guatemalan shadehouse will be different from culture under Dutch glass; yet, many of the same principles apply to both situations.

Nutrition

As owner of an agricultural laboratory in Florida, I deal with nutrition of foliage plants on a daily basis. There are different ways to fertilize foliage plants, and their success or failure depends largely on whether the nutritional requirements of the crop are met. The nutritional disorders commonly encountered in the various varieties are outlined. Many nutritional disorders in foliage, however, have not been observed or described.

Also, I have not given fertility rates in pounds per acre or kilograms per hectare per year.

Such information is really a waste of time for growers, and the practice of making fertilizer recommendations that way should be discontinued. Growers are interested in knowing what kind of fertilizer works and how much goes in the pot or the injector.

Leaf analysis rating standards are also provided for many of the crops. These rating standards were largely developed by current and former staff members at A & L Southern Agricultural Laboratories, Pompano Beach, Florida, which probably has as much experience with leaf analysis and foliage plants as any facility in the world. Rating standards for some foliage plants have yet to be established, so in some cases no leaf analysis rating standards are listed.

Diseases

People who grow or maintain tropical foliage plants have to contend with a fairly wide array of diseases, as the warm, moist conditions favored by most foliage plants are also conducive to the proliferation of pathogens. The diseases you are most likely to encounter with each variety are described in detail. Examples of many foliage plant diseases are also illustrated in the color plate section. I would like to remind you that knowing exactly what disease you are fighting is critical. Too many growers will spend hundreds of dollars on fungicides before they will spend five cents to find out what disease they are trying to control. The best growers utilize the disease diagnosis services available to them in order to know exactly what diseases they are dealing with and how to fight them.

Remember that to have plant disease, you must have the host plant, the pathogen, and the environmental conditions necessary for disease to develop. Changing any of the three will usually eliminate a disease problem. In most cases, both cultural and chemical control measures are listed, including those which are likely to be most effective. The chemicals I have recommended are registered in the United States at the time of this writing, but keep in mind that chemical labels change. Fungicides available in other countries may be sold under different names or may be different entirely. To protect yourself as well as your plants, read and follow the chemical labels precisely and don't apply anything that is not specifically registered for what you are growing.

INSECT AND MITE PESTS

Growers, interiorscapers, and homeowners all have to contend with insect infestations in foliage plants from time to time. Pressure from insects is probably greater in the tropics than anywhere else. I have outlined the insects you are likely to encounter for each variety, as well as their symptoms and control measures. As with disease control, chemicals for insect control must be specifically labeled for your situation. Pesticides registered for shadehouse use may or may not be registered for use in greenhouses or interiorscapes. Read the labels, communicate with the manufacturers, and talk to consultants or extension personnel.

I have also tried to indicate cultural control measures and integrated pest management techniques where applicable. IPM for foliage plants is still in its infancy, and my experience with it is limited. Where biological controls and IPM techniques work, by all means try to use them. The control measures I have outlined here are well proven and effective, though they are not the only ones.

DISORDERS

If you are going to have a dog, you are going to have fleas. Foliage plant production is similar in that every variety will have six or eight or 10 things that could possibly go wrong. Knowing what causes the various plant disorders and how to deal with them is what separates the good grower or interiorscaper from the average ones (if this were easy, we wouldn't need professionals, right?). Many of the disorders you are likely to encounter are listed, and examples of some of them can be found in the color plate section.

TRICKS

I don't believe I have ever seen a plant book with a "Tricks" section. Sometimes one little trick or piece of knowledge can make the difference between success and failure. This section simply contains tricks and techniques I have observed or discovered over the years, as well as little-known findings from scientific literature. In some cases I have used this section to reiterate the most important cultural factors required for success. This was my favorite section in the book, and I hope you find it useful.

INTERIOR CARE

This section is intended primarily for the interiorscaper, but for the home-owner and hobbyist, as well. The grower needs to realize that success beyond the initial sale is intertwined with the ability of the end user to maintain the product successfully. If a plant dies, the user is not likely to run out and get another one. Conversely, many plant sales are lost because the potential buy-ers are afraid plants will not do well for them. I hope that this section, along with the rest of the informaton, will give those who maintain foliage plants new insight into keeping their plants healthy and happy.

AGLAONEMA

Chinese Evergreen

HABITAT

The common name for *Aglaonema* is Chinese evergreen, as *Aglaonema modestum* was first cultivated by the Chinese for centuries before culture of these plants spread to Europe and the Americas. The common name is rarely used by *Aglaonema* growers today, however. The scientific name comes from the Greek *aglos* (bright) and *nema* (thread).

Most of today's commercial hybrids can be traced back to tropical Southeast Asia, specifically Malaysia, the Philippines, and Thailand. Aglaonemas grow vigorously in that part of the world, generally under dense shade trees. They tend to be found in the lower elevations, though some are found on forested slopes. Aglaonemas do better in their native habitat closer to swamps or in wet areas at the bases of mountain slopes near the coastline, which probably explains the plants' penchant for moisture and humidity as well as their general lack of cold tolerance. The variegation patterns probably evolved to help the plants absorb both the weak light of the forest floor as well as occasional flecks of direct sunlight.

USES

Aglaonemas are produced commercially in most parts of the world as freestanding specimens. Pot sizes generally range from 4 to 10 inches (10 to 25 cm), though other sizes are occasionally produced. Most of the time three to six cuttings per pot are placed close together in the container. Three 6-inch (15-cm) pots are sometimes placed in 14- or 17-inch (35- to 42.5-cm) pots for large interiorscape specimen production. Aglaonema cuttings are sometimes included in dish gardens. Hanging basket production is rather rare. Six- to 10-inch (15- to 25-cm) plants are frequently used in interiorscape plantings. Aglaonemas excel in interiors because of tolerance of low light conditions. Because of cold sensitivity, landscape plantings of aglaonemas are limited to tropical areas.

VARIETIES

There are about 50 *Aglaonema* species, plus countless hybrids and cultivars. Nomenclature of species and hybrids is quite jumbled and confused, although there have

Fig. 1 *Aglaonema* Silver Queen and Maria grown in Boynton Beach, Florida. (Courtesy of Marshall Horsman)

been efforts to organize and standardize *Aglaonema* nomenclature. The confusion may continue for some time because hybrids tend to have different names in different parts of the world, and the same or very similar hybrids may have several names even within a localized area. For example, I have heard *Aglaonema* Maria called Emerald Beauty, Silver Duke, and Esmerelda. Many new varieties come from public and private hybridization efforts, while others have resulted from sports, and still others have been found in the wild. Hybrids with differing characteristics can result from the same cross; therefore, hybrids produced independently may have slightly different characteristics as well as different names. The following are some of the more common varieties produced today.

Silver Queen is probably the most widely produced *Aglaonema* cultivar in the world. It is said to have originated from a cross between *A. curtisii* and *A. treubii* made by Nat DeLeon and then proliferated by Bob McColley at his Bamboo Nurseries in the early 1960s. The plant has grayish green leaves with white and silver to gray blotches. Silver Queen plants may vary a little in the variegation characteristics and in leaf size, presumably the result of separate crosses from the same parents. Silver Queen suckers fairly well, though it is rather cold sensitive (58F, 14.4C) and susceptible to bent tip. As winter progresses, plants become hardier and may tolerate 45F (7.2C).

Emerald Beauty is another commonly used name for the *Aglaonema* Maria variety. It was originally collected in the wild on Palawan Island in the Philippines. It is more compact and darker than Silver Queen, with deep green foliage highlighted by bands and flecks of silvery green. Maria, or Emerald Beauty, grows a little more slowly than Silver Queen and is rarely seen in pots larger than 10 inches (25 cm). Because of its fairly short pot habit, it tends to be more attractive in azalea pots or bulb pans, as opposed to full-sized containers. Emerald Beauty is more cold tolerant at 45F (7.2C).

Promising new cultivars include Maria Christina as well as the patented Elite series, including Queen of Siam, Rhapsody in Green, Amelia, and Green Majesty. Numerous older varieties exist that have good characteristics, though they are less commonly produced today.

PROPAGATION

Commercial aglaonema growers almost exclusively propagate from cuttings, either unrooted, callused, or rooted cuttings. Callused and rooted cuttings are a little faster and are sometimes less disease prone, although they cost more. Some varieties come true from seed, though, and plants can also be propagated by division and from stem pieces.

A cutting should generally have about five leaves. If the cutting is too big, it may flop in the pot and have a stretched appearance later on. A cutting that is too small with no stem tends to have disease problems and rot. The cuttings will sucker better if they are spaced in the pot, as opposed to being potted in a clump, though both methods yield a marketable plant. Rooting time is about four weeks in warm soil, longer in cooler soil.

Growers occasionally mist the cuttings during the warm part of the day, but excessive mist can lead to disease problems. Rooting hormones have given inconsistent results, and many growers don't use them on this plant. Cuttings are generally purchased through brokers, with most cuttings coming from ground beds in Central America, the Caribbean, Africa, and Southeast Asia.

CULTURE

Aglaonemas are generally not very picky about their soil mix. Various combinations of peat, bark, wood chips, sawdust, sand, and perlite are used successfully. As long as the mix has decent aeration and moisture-holding capacity, aglaonemas tend to grow pretty well. The pH target is generally 5.5 to 6.5, though they can tolerate lower pH. Growers frequently drench shortly after potting with a fungicide combination containing thiophanate methyl (Cleary's 3336, Domain) plus Subdue (metalaxyl), Aliette (fosetyl-aluminum), or Truban (etridiazole) for water mold fungi. Banrot (etridiazol + thiophanate methyl) is also registered. High-phosphate, soluble starter fertilizers tend to help expand the root system once cuttings have rooted.

The plant favors moist conditions and is a fairly heavy feeder, though not a particularly fast grower. Six-inch (15-cm) pots with three to five cuttings per pot are generally grown on 8- to 10-inch centers, usually on ground cover, sometimes on a bench, although benches are not essential for this plant. It normally takes about six to eight months to finish a 6-inch Silver Queen, depending on cutting size and the number of cuttings used. Maria (Emerald Beauty) tends to take a little longer. Larger containers are either planted with cuttings directly or are made by repotting one or more 6-inch (15-cm) plants into the larger container. Best light levels are 1,500 to 2,500 f.c. (16.1 to 26.9 klux), or around 80% shade in the tropics. When light is too high, foliage stays rather vertical, with pale color and tan blotches near leaf tips. When light is too low, the plant looks good but grows very slowly.

 You've been taught that plants take in carbon dioxide and give off oxygen, but actually at night the opposite happens.

NUTRITION

Most aglaonema varieties are relatively heavy feeders. They have somewhat high requirements for potassium, magnesium, iron, and copper. Aglaonemas can be grown with granular fertilizer as a topdress; with coated, slow-release fertilizers applied either as a topdress or incorporated into the soil mix; on constant liquid or soluble fertilizer; or with a combination of all of these methods. They do well with a 2-1-2 ratio of nitrogen-phosphate-potash, frequently with supplemental magnesium.

Nitrogen deficiency gives you small, pale leaves and little growth. Phosphorus deficiency is rather rare, usually resulting in only a weak root system. Potassium-deficient plants show necrosis in the older foliage, and the plant tends to shed older leaves. Magnesium deficiency is common, with broad yellowing of the margins of the older leaves. It is especially common in the darker varieties, which have more chlorophyll. Plants low in calcium have a thinner, softer leaf blade and weak foliage overall. Lack of iron shows the typical interveinal chlorosis in the new foliage. Copper deficiency is quite rare today because of copper fungicides and the evolution toward new varieties. Lack of copper cripples the new leaf, resulting in a very small, barely formed

TABLE 1 Leaf analysis rating standards for *Aglaonema commutatum*

Nutrient (%)	Very low	Low	Medium	High	Very high
Nitrogen	<2.3	2.3–2.6	2.7–3.5	3.6–4.1	>4.1
Sulfur	<0.15	0.15–0.19	0.20–0.75	0.76–1.00	>1.00
Phosphorus	<0.14	0.15–0.19	0.20–0.75	0.76–1.00	>1.00
Potassium	<2.25	2.25–2.69	2.70–5.00	5.01–5.99	>5.99
Magnesium	<0.20	0.20–0.29	0.30–0.60	0.61–0.80	>0.80
Calcium	<0.60	0.60–0.99	1.00–2.00	2.01–2.75	>2.75
Sodium			0.11	0.11–0.20	>0.20
(ppm)					
Iron	<30	30–49	50–300	301–1000	>1000
Aluminum			50–250	251–2000	>2000
Manganese	<30	30–49	50–300	301–1000	>1000
Boron	<16	16–24	25–50	51–75	>75
Copper	<6	6–9	10–100	101–200	>200
Zinc	<16	16–24	25–200	201–1000	>1000

Sources: Institute of Food and Agricultural Sciences, Apopka, Florida; Dr. Benjamin Wolf, Fort Lauderdale, Florida.
Notes: Common names include Chinese evergreen, Silver Queen, Pseudobracteatum, Treubii, Fransher, Maria. Sample of most recent fully mature leaves, no petioles.

leaf. Boron toxicity is manifested as tan to brown blotches just back from the leaf tip on the larger half of the leaf blade.

DISEASES

The most serious diseases of aglaonemas are *Erwinia* bacteria, both *E. carotovora* and *E. chrysanthemi*. *Erwinia* causes bacterial soft rot of either stem or leaf tissue. Its symptoms are a wet, slimy rot of leaves or stems, sometimes but not always associated with a foul odor. It is not the bacterium that causes the odor but the rapid breakdown of the tissue. Problems with *Erwinia* are common during propagation and somewhat less after the plants are rooted. *Erwinia* stem rot shows a wet, mushy rot with yellowing of foliage. Cultural controls are the most effective; use clean cuttings and keep a clean greenhouse. Growers commonly use copper fungicides, frequently mixed with Dithane (mancozeb). Agrimycin 17 (streptomycin sulfate) is also used, but it is not labeled for *Aglaonema* in the United States. Sprays of vinegar solutions also seem to help.

Fusarium stem rot, another common disease, frequently originates in stock plants. It is a drier rot than *Erwinia,* with the internal stem tissue appearing white and somewhat mealy inside. The edge of the infected stem tissue frequently has a purple or red appearance. Thiophanate methyl fungicides are generally used against *Fusarium* stem rot, sometimes in combination with Captan. Keeping the soil pH up may help, as well. *Pythium,* a common root rot fungus of *Aglaonema,* generally occurs under wet conditions or in heavy, poorly drained soils. Growers usually use Subdue (metalaxyl), Truban (etridiazole), or Aliette (fosetyl-aluminum) to combat *Pythium.*

Nematodes are a major problem in *Aglaonema,* specifically lesion *(Pratylenchus)* and root knot *(Meloidogyne).* I have consulted for aglaonema stock farms throughout the Caribbean and Central America and in parts of Africa, and I have never seen an aglaonema cutting farm that was not infested with nematodes. Stock plant growers generally treat two to three times a year with a nematicide, using such items as Counter (terbufos), Furadan (carbofuran), Mocap (ethoprop), Temik (aldicarb), and Vydate (oxamyl), though it is unlikely any of these is registered for their situations. The symptoms are a hollowing out of the roots, where only the root cortex remains, with the roots resembling a drinking straw. Plants lose vigor and leaf size, and cuttings fall over. The plant also tries to send out new roots higher up on the stem. No chemical nematicides for foliage are currently registered in the United States, which is a travesty. Nematodes can come in on unrooted or rooted cuttings, and once the population builds up, the plants decline later in life. Hot-water cutting dips at 122F (50C) deserve further investigation.

The most common leaf spots of *Aglaonema* are *Myrothecium* and *Colletotrichum. Myrothecium* frequently attacks wounded tissue, especially in propagation. It is a large, dry, brown leaf spot, usually circular, with visible white and black fruiting bodies on the foliage underside. Removing affected leaves is helpful, and sprays of Dithane (mancozeb) or Daconil (chlorothalonil) are common. *Colletotrichum* is similar to *Myrothecium,* but it tends to spread out more, becoming more of a leaf blight. Sprays for *Colletotrichum* are similar to those for *Myrothecium,* though copper fungicides are also used.

Two less common diseases are *Xanthomonas* and dasheen mosaic virus. Plants infected with *Xanthomonas* show brownish and yellowish tissue along the margins and leaf tips, especially in older leaves. The symptoms look more like a burn or toxicity rather than a disease. The lesions are not water soaked, and there are no fruiting bodies. The bacterium is spread primarily by splashing water, and the most effective control is to remove infected leaves with sterilized tools. Sprays of copper and Dithane (mancozeb) are also used, as is Phyton 27 (picro cupric ammonium formate) or Aliette (fosetyl-aluminum).

Dasheen mosaic virus is rare today. You get a mosaic, or blotching, symptom on the foliage, and the plants tend to be stunted, with some leaf distortion. It spreads primarily by cutting tools, though aphids can spread it. Don't waste your time with virus-infected stock. Throw it away, sterilize, and start again from a different cutting source.

INSECT AND MITE PESTS

Fortunately, aglaonemas have few insect pests. Foliar and root mealybugs, the most common ones, are small, white, segmented insects, frequently with red juice inside them. Growers usually use Diazinon or Dycarb (bendiocarb) for foliar mealybugs. Talstar (bifenthrin) and Cygon (dimethoate) are also frequently used. Root mealybugs are usually controlled with Diazinon drenches.

Problems with scales, aphids, and mites are very rare with this plant. Thrips can be a problem at times. They work within the tube of the unfurled leaf, giving you a leaf that tends to be torn on one side, with numerous small, brown blotches similar to a disease symptom. Sprays of Mavrik (tau-Fluvalinate), Orthene (acephate), or Avid (abamectin) are common, especially with a wetting agent or penetrant. Several sprays in rotation at three- to four-day intervals are usually necessary.

An occasional problem that looks like an insect problem, but is not, is the bird's-nest fungus, also known as shotgun fungus or glebal masses. This has been known since 1957. Looking like a scale insect, as small, brown disks on the foliage underside, it is caused by the fungus *Sphaerobolus* and similar fungi. The fungus is actually soilborne, but its fruiting bodies show up on the foliage. It is definitely not a scale insect, and it is normally controlled by removing the affected leaves and using clean soil. Captan sprays seem to be helpful.

DISORDERS

Bent tip is a common disorder of aglaonemas. It affects mostly Silver Queen, but also a few other varieties as well. It apparently does not have a nutritional cause. High light levels and water stress have been implicated, though personally, I don't believe it. It is my experience that when aglaonemas grow too fast, the tip of the leaf catches on itself early in the unfurling process, giving you a bent tip symptom. It is not a big production problem and is fairly rare under interior or homeowner's conditions, where growth is slower.

Tip burn is a common sympton when soluble salts are excessive. You get a dry, tan or brown tip on the edge of the leaf, symmetrically going across the leaf tip. Boron toxicity tends to be asymmetrical and to occur on only the bigger half of the leaf tip. Fluoride toxicity is reported, but it is fairly rare in aglaonemas. Keep the soil pH and calcium levels up to reduce fluoride uptake.

Cold injury is rather distinctive on aglaonemas, usually resulting in a dark, greasy appearance on the upper leaf surface. It is the result of epidermal collapse and tends to affect older leaves as well as older plants.

TRICKS

Leaf size and thickness can be increased very significantly with calcium nitrate sprays, using about 2^1/$_2$ pounds per 100 gallons (1.1 kg/400 l). Chelated calcium can also be used. Most chelated iron products work as a drench in controlling iron deficiency in this plant, but in my experience only the EDDHA chelates, such as Sequestrene 138, work as foliar sprays. Aglaonema cuttings can be rooted in water, and this practice is commercially viable. In Puerto Rico I have seen the plant rooted in plastic cups with about one-half inch (1.25 cm) of water in the bottom. In Central America I have seen the plant rooted by placing the cutting in a black plastic sheath. Sprays of benzyladenine will cause aglaonemas to sucker profusely, although I have not seen this done on a commerical basis.

Bonzi has been used to control height in large, stretchy varieties, such as Abidjan. Aglaonema roots do not like light, and root systems tend to grow better in black or green pots rather than opaque white pots. The roots locate more in the internal part of the medium when growing in an opaque pot. Also, sprays of 250 ppm gibberellic acid induce flowering for breeding purposes.

INTERIOR CARE

Aglaonemas are excellent, attractive houseplants that last for years indoors as long as they get just a little bit of care. Minimum light level is 100 f.c. (1.1 klux), though 150 to 250 f.c. (1.6 to 2.7 klux) are preferred. Best temperatures are 70 to 75F (21 to 24C). Do not let the temperature fall below 60F (16C) for Silver Queen. Most of the other varieties will tolerate lower temperatures.

If aglaonemas dry out excessively, the older leaves will begin to turn a solid canary yellow, then brown. Irrigate about once a week. Fertilize about every three months with 1 teaspoon of 20-20-20 per gallon (3.8 l), because aglaonemas continue to grow in the interior environment. Be aware that sometimes nematode populations can cause severe root problems if you purchase infested plants.

REFERENCES

Birchfield, W., J.L. Smith, A. Martinez, and E.P. Matherly. 1957. Chinese Evergreen Plants Rejected Because of Glebal Masses of *Sphaerobolus stellatus* on Foliage. *Plant Disease Reporter* 41 (6): 537-539.

Brown, B. Frank. May 1996. Telephone conversation with the author.

Chase, A.R. 1993. Common Diseases of Aglaonema. *Southern Nursery Digest* (April): 20-21.

Conover, C.A., R.T. Poole, R.J. Henny, R.A. Hamlen, and A.R. Chase. 1981. *Aglaonema Production Guide for Commercial Growers*. IFAS OHC-1. Gainesville: University of Florida IFAS.

Fooshee, W.C., and McConnell, D.B. 1987. Response of *Aglaonema* 'Silver Queen' to Nighttime Chilling Temperatures. *Hortscience* 22 (2): 254-255.

Griffith, L.G. 1982. Common Problems With Aglaonemas. *Florida Nurseryman* (September): 10, 69.

Poole, R.T., and Pate, A.J. 1980. Bent-tip of Aglaonema commutatum 'Silver Queen'. *Foliage Digest* (September): 10.

Simone, G.W. The Scale Problem That Isn't. 1985. *Interior Landscape Industry* (March): 60-66.

ALII

(See Ficus maclellandii*)*

ANTHURIUM

HABITAT

The genus name *Anthurium* is derived from two Greek words meaning "tail flower." There are about 600 species of *Anthurium*, about 500 native to Colombia and the rest to adjacent areas of tropical America. Like orchids, anthuriums can be either terrestrial or epiphytic. Most come from wet, humid areas in northern South America.

USES

Anthuriums are desirable plants not only for their colorful, long-lasting, tropical flowers but also for their deep green, shiny, arrow-shaped leaves. *Anthurium* is one of the better low-light plants for home and interior use because it can continue to flower with relatively little light,

Fig. 2   *Anthurium* Lady Beth grown from tissue culture. (Photo by the author)

making it a true flowering foliage plant. Individual blooms can last as long as seven or eight weeks.

With the introduction of suitable varieties, it has become a major tropical foliage item only in the last 10 years. Most production as foliage plants is for 6-, 8-, and 10-inch (15-, 20-, and 25-cm) pots. Azalea pots are frequently used instead of full-sized containers because of the relatively compact nature of the most popular varieties. Anthuriums are also grown for cut flowers in Jamaica, Trinidad, Hawaii, and elsewhere.

VARIETIES

Most of the cut flower as well as foliage varieties today have been derived from crosses involving *A. andraeanum*, a Colombian native, and *A. scherzeranum* (frequently misspelled *scherzerianum*), which is native to Guatemala and Costa Rica. Anthuriums were a somewhat minor foliage item until Oglesby Nursery developed the Lady series of dwarf, tissue-culture anthuriums. Lady Jane was the first major introduction, a sturdy, dwarf, red-flowered variety that is probably still the best of the series. Two other varieties from this series are Lady Anne, similar to Lady Jane but with white flowers, and Lady Beth, a pastel pink variety. Another interesting type is the Kohara Double, a true double red anthurium, which can be a challenge to grow under hot summer conditions.

A number of *Anthurium* varieties are grown primarily for their attractive foliage patterns, as opposed to their flowers. *A. crystallinum*, known as Crystal or Crystal Hope anthurium, has velvety green leaves with contrasting silver veins, looking somewhat like a calathea in leaf pattern. It too can be difficult in heat. Also, many current breeding efforts involve *Anthurium amnicola* as a parent, and the resulting hybrids may require less fertilizer.

PROPAGATION

All common forms of propagation can be used successfully on anthuriums, including cuttings, division, and seed production. However, today the overwhelming majority are produced from tissue culture. Providing not only reasonably rapid proliferation of a new variety, tissue culture also helps keep the plants free of certain bacterial and viral diseases. Cloned plants also flower more quickly than seed varieties. Starter plants are generally sold as Stage 4 liners in cell trays. Liners should be removed from trays very carefully, as they are rather delicate, and rough handling will lengthen production time.

CULTURE

Anthuriums grow best in media with very good drainage and somewhat lower moisture-holding capacities than average foliage plant media. They like various wood products in the mix, such as pine bark or cypress chips. Mixes very heavy in peat tend to hold too much moisture and contribute to root problems. A 1-1-1 ratio of sphagnum peat, composted bark, and perlite works well. I have seen anthuriums grown as cut flowers in the humid tropics using nothing but pieces of brick, coconut husk, or railroad stone as the growing medium.

Typical light levels are generally between 1,500 and 2,500 f.c. (16.1 and 26.9 klux), though there are some varietal preferences within that range. It helps to grow the plant toward the bright side early, as it tends to be sturdier and a little more compact, with better suckering. When grown too bright, the plant tends to be pale and somewhat bleached, with occasional foliar burn. When grown too dark, the leaf petioles are quite long, giving the plant a somewhat stretched, upright appearance with fewer flowers. The most favorable temperature range is 65 to 85F (18.3 to 29.4C), with night temperatures in the 60s to lower 70s F (17 to 23C). The plant will tolerate 45F (7.2C), but tends to suffer cold injury in the low 40s (4.4C). Cold injury symptoms include reddening, especially of the older foliage, and sometimes marginal burn of the leaves. A little air movement is nice, but the plant hates windy conditions.

NUTRITION

Anthuriums are not particularly heavy feeders. Early publications recommended the equivalent of 1,200 pounds of nitrogen per acre per year, but later research tends to indicate a lower requirement. They perform best with a slow, steady supply of nutrients rather than a large dose of fertilizer at once. You cannot push anthuriums, as they grow pretty much at their own speed. The plants tend to do a little better with higher fertilization rates when grown under higher light, with less fertilizer when grown under lower light.

The most common fertilization method is constant liquid feed with 200 ppm nitrogen (not over 250) of a 3-1-2 ratio of $N-P_2O_5-K_2O$, although research shows a 1-2-2 ratio is preferred. They are also grown with long-term, coated, slow-release fertilizers either top-dressed or incorporated into the media. Combinations of slow-release and soluble fertilizers also work well, but top-dressings with granular fertilizers are discouraged.

Plants deficient in nitrogen are characterized by pale foliage, with yellowed older leaves and stretched, very droopy petioles. When low in phosphorus, the plant is stunted and has small leaves. Potassium deficiency also gives you stunting and yellow older foliage, with yellow leaf margins and necrotic leaf spots especially throughout the older leaves. Anthuriums may have the highest magnesium requirement of any foliage plant, though a relatively low demand for trace elements. Lack of magnesium results in chlorotic margins in the older foliage, and in some varieties there are chlorotic bands moving in toward the midrib. Magnesium requirements are higher in bright light. Sulfur deficiency gives you a slight stunting and a somewhat pea green color rather than a deep green. With lack of calcium, the flower dies, and you can get a short, stubby spadix. Sometimes lack of calcium results in a somewhat cupped and deformed spathe.

DISEASES

Much has been written about the primary disease of this plant, *Xanthomonas campestris* pv. *dieffenbachiae*, first reported in Brazil in 1960. This disease has been a problem for many years in cut-flower anthuriums, causing leaf spot and blight, as well as systemic vascular collapse. With the advent of tissue culture and improved cultural controls, *Xanthomonas* is much less of a problem. Chemicals such as Agrimycin 17 (streptomycin sulfate), the copper fungicide Phyton 27 (picro cupric ammonium formate), and Aliette (fosetyl-aluminum), offer a limited degree of control, but the best controls are cultural. Sanitation and sterile cutting tools are very important. Try not to work in anthurium greenhouses while the foliage is wet. Irrigate toward the middle of the day, when drying conditions are good. Fertilizer rate and source seem to have minimal influence.

We use Latin names for plants
because Latin is a dead language and will
therefore never change.

Root knot and burrowing nematodes are significant problems in cut flower beds, but these problems are rare with tissue-culture plants grown in commercially prepared media. Plants kept too wet frequently develop *Pythium* and *Rhizoctonia* root rot, which can be somewhat controlled with Subdue (metalaxyl) or Truban (etridiazole) for *Pythium,* Terraclor (PCNB) or thiophanate methyl for *Rhizoctonia.*

INSECT AND MITE PESTS

The most troublesome pests by far are spider mites and thrips. Mite activity shows up as small, white specks on the foliage, sometimes with a mild cupping or distortion of the leaf. Thrips rasp on foliage, giving a more distorted appearance, with numerous small, brown, irregularly sized spots. Thrips also frequently feed on unfurling blooms, causing tan to brown spots on the flower. Growers frequently use Pentac (dienoclor), Avid (abamectin), Mavrik (tau-Fluvalinate), and other miticides. Thrips are frequently controlled with Mavrik, Thiodan (endosulfan), and Tempo (cyfluthrin).

Problems with snails and slugs are common when growing on ground cover in moist greenhouses. Symptoms include numerous holes in the foliage and slime trails. Metaldehyde baits help somewhat. Mesurol or Grandslam are also effective but are not available in some markets. Whiteflies are occasionally a problem with young plants, and Talstar (bifenthrin) can be helpful.

Disorders

Edema is sometimes observed as numerous small, brown, protruding bumps on the older foliage. This is caused by moisture level fluctuations, resulting in the leaves temporarily containing too much water. Try to avoid wide fluctuations of moisture levels when temperatures are high.

Orthene (acephate) has a history of phytotoxicity on anthuriums, causing distortion of leaves and flowers. When copper fungicides are used too frequently, plants develop a rather hard, somewhat brittle appearance. Anthuriums are sensitive to high salts, frequently resulting in leaf tip burn. Marginal scorch of older leaves can occur when salts are high along with excessive medium drying.

Tricks

Despite the high magnesium requirement of anthuriums, foliar magnesium sprays are only somewhat effective on this plant due to the rather thick, waxy leaf cuticle, which reduces foliar absorption. Magnesium is better applied through liquid feed or soil drenches. Try to maintain a calcium-to-magnesium ratio of less than 8 to 1 in the soil. Nematodes can be controlled in cuttings with hot water baths at 122F (50C) for 10 minutes. Suckering can be induced with a spray of 1,000 ppm benzyladenine. Blooms from Lady Jane make excellent cut flowers, especially when you put some floral preservative or soft drink in the vase. When using Aliette (fosetyl-aluminum) for disease control, do not spray less than two weeks before or two weeks after a spray containing copper, or phytotoxicity could result. Some varieties may be sensitive to Dybarb (bendiocarb).

INTERIOR CARE

If given sufficient light indoors, anthuriums may continue to grow and flower. Try to give them at least 500 f.c. (5.4 klux) or place them near a window. Severe root rot may develop if plants are kept too wet, so let the medium dry before irrigating. Try to maintain at least 40% relative humidity, if possible.

Anthuriums are susceptible to spider mites and mealybugs indoors, so inspect for pests periodically. Because of their sensitivity to soluble salts, be very conservative when fertilizing anthuriums. One-half teaspoon of soluble 20-20-20 per gallon (3.8 l) every three months or so should be sufficient indoors. It also helps to add Epsom salts at about one-half teaspoon per gallon along with the 20-20-20. Try to keep temperatures above 50F (10C), and avoid cold or drafty areas.

REFERENCES

Akamine, E.K., and T. Goo. 1975. Vase Life Extension of Anthurium Flowers with Commercial Floral Preservative, Chemical Compounds and Other Materials. *Florist's Review* (January 30): 14-15, 56-59.

Henny, R.J., and W.C. Fooshee. 1989. Response of *Anthurium* 'Lady Jane' Liners to Different Light and Fertilizer Levels. *Foliage Digest* (October): 7-8.

Higaki, T., and H.P. Rasmussen. 1979. Chemical Induction of Adventitious Shoots in *Anthurium*. Hortscience 14 (1): 64-65.

Hotchkiss, S. 1992. Assorted Beauties. *Greenhouse Grower* (September): 112.

Hotchkiss, S. 1995. Anthurium. *Greenhouse Management and Production* (July): 8.

Imamura, J.S., and T. Higaki. 1984. *Nutrient Deficiency in Anthuriums*. University Of Hawaii Research Extension Series 047. Honolulu: College of Tropical Agriculture and Human Resources.

Imamura, J.S., and T. Higaki. 1988. Effect of GA and Ba on Lateral Shoot Production on *Anthurium*. Hortscience 23 (2): 353-354.

McConnell, D.B., and R.W. Henley. 1987. Plant Profile for Interiorscapes: Crystal Anthurium. *Foliage Digest* (June): 7.

APHELANDRA
Zebra Plant

HABITAT

Commonly known as zebra plants, most aphelandras are native to tropical America, especially Brazil. In the wild they can get up to 2 feet (61 cm) tall, but most of the commercial cultivars today are dwarf varieties chosen for their compact pot habits. The plants are also sometimes called saffron spikes, due to their predominantly yellow flowers. They are generally found under rather heavy shade, as an understory plant, in the wild.

Fig. 3 Gorgeous aphelandras in an old Apopka, Florida, greenhouse. Note the fans and heater for climate control. (Courtesy of Marshall Horsman)

USES

Aphelandra is an excellent low-light interior plant and is another example of a true flowering foliage plant. Even without flowers, its multicolored foliage is quite attractive. Today the plant is generally grown in 4-, 5-, and 6-inch (10-, 13-, and 15-cm) pots. Single-stem production is most common, although three plants per pot are sometimes produced in larger containers, as are branched cuttings. One plant per pot well spaced seems to be the most common production scheme. When watered correctly the zebra plant is a good, colorful, low-light plant that can bloom indoors.

VARIETIES

Historically, there have been four principal varieties of *Aphelandra*, two mostly dark green and two more variegated. Dania and Louisae are compact, dark green varieties with bright white leaf veins. Both display bright, predominately yellow bloom spikes. Dania is probably the most widely grown aphelandra. Apollo and Red Apollo are more variegated types, with the upper leaf surfaces being largely whitish or ivory in color. Red Apollo has more maroon in the stem and on the lower leaf surfaces. Silver Cloud is another attractive variety. All five varieties are cultivars of *Aphelandra squarrosa* Nees.

PROPAGATION

Almost all propagation is done from cuttings, though in various ways. Tip cuttings from stock plants appear to yield the best finished plants. Tip cuttings should have at least one set of full-sized leaves. Try to leave at least two sets of leaves on the stock plant to keep it vigorous. Stock plants will get tired after a time and will need to be replaced. Growers producing branched or multiple plants propagate from single-eye cuttings, similar to a pothos eye, or a double-eye cutting or joint, which is a cutting from middle stem tissue, back from the tip. Good stem caliper is important for any aphelandra cutting.

Rooting hormones help a little bit, and mist every 30 minutes or so is critical. Get the cuttings stuck under mist quickly to avoid excess wilt, which happens fast. Cuttings root in about three weeks under warm summer conditions, up to six weeks in cool winter conditions. The cuttings seem to do better if they have whole leaves. Bottom heat is helpful in the cold climates.

The pH scale is logarithmic, just like the Richter scale.

CULTURE

A forgiving potting mix is critical for aphelandra production. It needs to hold some moisture but have good aeration at the same time, as this plant requires regular, steady watering but will die quickly when the mix is waterlogged. Combinations of peat, bark, and wood chips are used, as are peat, bark, and perlite mixtures. The pH is usually adjusted to 5.5 to 6.5. Plants need enough space to avoid having the leaves touch each other. Six-inch (15-cm) pots are usually spaced 8 by 8 inches (20 by 20 cm). Don't let the foliage overlap.

Aphelandras are not photoperiod sensitive, but light intensity, duration, and temperature all influence flowering. Light levels for production range from 800 to 1,500 f.c. (8.6 to 16.1 klux). Under low light the leaves are long, narrow, and flat. Under higher light conditions the leaves are shorter and wider, with more puckering and more distinct white variegation. Ideal daytime temperature is 80F (27C), with a minimum desired temperature of 65F (18C) and a maximum of 90F (32C). Optimum soil temperatures are 70 to 80F (21 to 27C). Growing aphelandras cool tends to delay flowering.

Irrigate carefully, trying to avoid extremes of wet and dry. Drip irrigation is common, as are hand watering, capillary mats, and ebb-and-flow systems. Bench production is preferred. Single-cutting 4-inch (10-cm) plants finish in flower in 15 to 20 weeks, whereas single 6-inch (15-cm) plants finish with a flower spike in 18 to 22 weeks.

NUTRITION

Aphelandras are moderate feeders. Constant liquid feed is the most common type of fertilizer application, most commonly consisting of 200 ppm nitrogen, 75 ppm phosphorus, and 150 ppm potassium. The plant tends to like iron and calcium.

Don't try to push *Aphelandra* too much with liquid feed, as this tends not to make it grow faster and may cause problems. Elevated soluble salts can give you leaf burn, especially when running a little dry. The combination of drought and high salts can also result in leaf drop, and since the plant doesn't have many leaves to begin with, leaf drop can cause a high percentage of culls. Too much fertilizer also increases possibilities for *Myrothecium* leaf spot and *Phytophthora* stem rot. Slow-release, coated fertilizers are also used on this plant at times, typically at 5 to 7 grams per pot.

Plants tend to be pale and grow slowly when they are under-fertilized. Iron deficiency, which gives a yellowing in the new foliage, can be a sign of root problems. It is usually best corrected with an iron drench.

DISEASES

Myrothecium is the most common foliar disease, causing a brown to black, roughly circular leaf spot. It especially happens on older leaves and leaves

---

TABLE 2 Leaf analysis rating standards for *Aphelandra squarrosa*

Nutrient (%)	Very low	Low	Medium	High	Very high
Nitrogen	<1.6	1.6–1.9	2.0–3.0	3.1–3.6	>3.6
Sulfur	<0.15	0.15–0.19	0.20–0.30	0.31–0.40	>0.40
Phosphorus	<0.15	0.15–0.19	0.20–0.40	0.41–0.60	>0.60
Potassium	<0.90	0.90–1.19	1.20–2.00	2.01–2.75	>2.75
Magnesium	<0.35	0.35–0.49	0.50–1.00	1.01–1.30	>1.30
Calcium	<0.30	0.30–0.59	0.60–2.00	2.01–2.50	>2.50
Sodium			<0.11	0.11–0.20	>0.20
(ppm)					
Iron	<30	30–49	50–300	301–1000	>1000
Aluminum			<251	251–2000	>2000
Manganese	<30	30–49	50–300	301–1000	>1000
Boron	<20	20–34	35–45	46–70	>70
Copper	<6	6–9	10–50	51–75	>75
Zinc	<15	15–19	20–200	201–1000	>1000

Sources: Institute of Food and Agricultural Sciences, Apopka, Florida; Dr. Benjamin Wolf, Fort Lauderdale, Florida.
Notes: Common name: zebra plant. Sample of most recent fully mature leaves, no petioles.

---

exposed to wounds, similar to aglaonemas. Apollo gets *Myrothecium* worse than Dania. Black and white fruiting bodies are visible on the underside of the spots, and infected leaves tend to drop. *Myrothecium* can also cause a cutting rot in propagation. Daconil (chlorothalonil) and Zyban (thiophanate methyl plus mancozeb) are effective controls.

Corynespora causes a similar leaf spot but without fruiting bodies. Chemical controls are the same as for *Myrothecium*. In cool winter weather *Botrytis* can cause foliar blight, especially when too much mist is applied to cuttings. Clean up all old leaf debris. Sprays of Chipco 26019 (iprodione) or the chlorothalonil formulations Daconil or Exotherm Termil are effective.

The most common stem and root rot is caused by *Phytophthora,* which starts as black lesions on the lower part of the stem and ends up as a mushy, black stem rot. It frequently happens when plants are too wet. Sprays or drenches of Aliette (fosetyl-aluminum) are effective, and the plant seems to like them. Subdue (metalaxyl) as a drench is also effective.

In hot, wet conditions *Rhizoctonia* root rot is common. It is distinguished by small, reddish brown threads on the soil surface, especially under leaves touching the medium. Drenches of Chipco 26019 (iprodione), Terraclor (PCNB), or Terraguard (triflumizole) are effective.

A rare stem gall disease is caused by the fungus *Kutilakesopsis*. Chemicals are generally not effective here, and plants with galls should be dumped.

INSECT AND MITE PESTS

While insect pests are not extremely common on zebra plants, mealybugs, aphids, thrips, broad mites, and scale insects do occur. When white, cottony, segmented mealybugs are seen, sprays of Diazinon or Cygon (dimethoate) are helpful. These sprays also control aphids. Thrips tend to rasp on young, unfurling leaves, giving a silvery, flecked appearance, sometimes with distortion. Their small, black droppings tend to be more visible than the thrips themselves. Mavrik (tau-Fluvalinate) and Avid (abamectin) are effective controls, especially in conjunction with a wetting agent.

Broad mites are too small to see without magnification, but they cause leaf curl, distortion, and downward cupping of foliage. Avid (abamectin), Kelthane (dicofol), and Pentac (dienochlor) are effective with multiple applications. The most common scale is hemisphaerical scale, a large, dark brown scale which secretes honeydew and contributes to sooty mold. Cygon (dimethoate) is a useful control, but spray when it's cool.

DISORDERS

By far the most common disorder is leaf crinkle, also called leaf stunt. Apparently healthy plants begin to show downward leaf curl, leaf twisting, and small foliage. It does not seem to be nutritional in nature, nor does it appear to be caused by a fungus, bacterium, or virus. Leaf crinkle seems to be systemic in the plant, as cuttings taken from crinkled plants tend to stay crinkled. High light and warm conditions exacerbate the problem. Avoid excessive light, and dump plants with leaf crinkle. One theory suggests that when aphelandras grow too fast, the green part outgrows the white part, and crinkle results.

Other common disorders include marginal burn of older leaves, which is usually related to elevated soluble salts. Water-stressed plants also drop leaves, especially if salts are elevated.

TRICKS

The main secrets to this plant are to get the soil mix right and the light levels right. If you can accomplish those two things, 90% of your problems are solved. Muriate of potash (KCl) can be used as a potassium source. The growth regulator A-rest (ancymidol) will shorten internodes and keep the plants short, though I have not seen it applied commercially. Aphelandras like air movement, which seems to give them larger leaves, so keep some fans in the greenhouse. Finally, don't pack the soil when you pot, as that tends to lead to disease problems later.

INTERIOR CARE

Aphelandras do well in interior conditions under moderate light, 150 to 200 f.c. (1.6 to 2.1 klux). Avoid exposure to direct sunlight. Try not to place them where they will be exposed to cold drafts. Irrigate regularly to keep them moist, but don't apply too much water at one time, because disease would likely follow. It is okay to let aphelandras wilt slightly prior to watering. Give them very little fertilizer, no more than a teaspoon of a typical 20-20-20 fertilizer per gallon every two to three months.

REFERENCES

Batson, F. 1973. Aphelandra, 2-Way Plant. *Florist's Review* (April 5): 28-29.

Chase, A.R. 1991. Effect of Osmocote Rate on Severity of Phytophthora Stem Rot of *Aphelandra squarrosa* Dania. *Foliage Digest* (January): 2-3.

Chase, A.R., and R.T. Poole. 1986. *Effect of Nutrition and Growth of* Aphelandra squarrosa *and Severity of Myrothecium Leaf Spot.* Arec-Apopka Research Report RH-86-3: 1-3. Apopka: University of Florida IFAS Agricultural Research and Education Center.

Henley, R.W. 1982. *Aphelandra Leaf Crinkle—Aphelandra Stunt: A Report From Florida.* ARC Research Report RH-82-1: 1-3. Apopka: University of Florida IFAS Agricultural Research Center.

Osborne, L.S., and A.R. Chase. 1990. Phytotoxocity Evaluations of Dycarb on Selected Foliage Plants. *Nursery Digest* (January): 30-31.

Poole, R.T., and C.A. Conover. 1982. Light Intensity Influences Flowering of Aphelandra. *Foliage Digest* (January): 13-14.

Poole, R.T., C.A. Conover, A.R. Chase, L.S. Osborne, R.J. Henny, and R.W. Henley. 1983. *Aphelandra Production Guide.* IFAS OHC-9: 1-6. Gainesville: University of Florida IFAS.

ARALIA/POLYSCIAS

HABITAT

The Araliaceae family contains numerous ornamental species, mostly shrubs and small trees. Many of them are native to Polynesia and other Pacific islands, though a few are native to North America. A number of the species used in foliage production today have been reclassified from the genus *Aralia* into the genus *Polyscias*. *Fatsia japonica*, which used to be called *Aralia sieboldii*, and *Dizygotheca elegantissima*, the false aralia, now are also in distinct genera.

Aralias and polysciases frequently are flowering shrubs in the wild, though blooms are fairly uncommon in cultivation. The plants are very commonly found as hedges in tropical areas, specifically USDA Zones 10 and 11. They generally have alternate leaves, and many variegated forms exist.

Fig. 4 *Aralia* Fabian, with broad, cupped leaves that are purple underneath. (Photo by the author)

USES

Polyscias varieties are most often grown in larger pots, from 6 to 14 inches (15 to 35 cm) and larger. They are usually produced as multiple cuttings per pot, though sometimes single-branched cuttings are used. Their upright pot habit makes them useful for narrow spaces in interior environments. The aralias make good interior plants with adequate light. Small cuttings are used in dish gardens, and they are sometimes planted as screens or windbreak hedges around nurseries.

VARIETIES

The nomenclature of aralias and polysciases was at one time a complete mess, with confusion not only among varietal names but in the genus and species names as well. Fortunately, several researchers have done a good job of cleaning up the nomenclature. The common name remains "aralia," but many of the commercially produced foliage plants are now in the genus *Polyscias*, though the old names persist.

The common Balfour aralia, now called *Polyscias pinnata,* has variegated green-and-white marbled foliage and an upright habit. The black aralia is *P. guilfoylei,* also called Blackie. It is a very attractive plant with quite dark foliage, hence the name. Chicken Gizzard aralia, *P. crispa,* has emerald green, crisp, curled leaves and a somewhat dwarf, upright appearance. The Dinner Plate aralia, also a cultivar of *P. pinnata,* has large, round leaves. Pennockii and Marginata are somewhat less common but attractive cultivars of *P. pinnata.*

Fabian aralia is a cultivar of *P. scutellaria.* The leaves are purplish on the underside, and it is therefore sometimes called Plum aralia. The oriental-looking Ming aralia is a larger variety of *P. fruticosa.* Parsley aralia is also a *P. fruticosa.* The leaves are somewhat smaller than Ming foliage, and they fairly closely resemble those of the parsley plant.

PROPAGATION

Cuttings are used almost exclusively to propagate members of the aralia family. Small, soft cuttings are used for some varieties, and larger cane pieces are also rooted. Cuttings are generally direct-stuck in the finishing container. Cutting size is frequently about 8 inches (20 cm), though both smaller and larger cuttings are used, depending on variety. Branched canes are produced by some Central American cutting farms. Even very large cuttings can be rooted fairly easily if conditions are right. Rooting hormone dips containing indolebutyric acid (IBA) are helpful.

The cuttings are misted occasionally during warm periods. It is best to operate the mist manually rather than from a clock; aralias don't need a lot of mist, and overmisting will almost certainly lead to disease problems. The plants prefer warm soil, about 75F (25C). It is important to get fresh cane cuttings, as the rooting percentage would be low if the cuttings were stored too long. Rooting time runs around three to four weeks, depending on temperature, variety, and cutting size.

CULTURE

Aralias are generally grown in higher light than many foliage plants. Production light levels range from 1,500 to 4,500 f.c. (16.1 to 48.4 klux). The variegated varieties tend to be grown brighter, and their color contrast is more intense under higher light. The solid green cultivars, such as Ming and Parsley, tend to be grown darker.

Aralias are not very tolerant of high moisture levels or flooding, so they need a rather open potting medium with a low moisture-holding capacity. Mixtures of peat and wood products, such as bark or cypress, are frequently used, and many growers like to have 10 or 15% sand in the mix to help keep plants from blowing over in shadehouses. As long as elevation is good, they can be grown well on ground cover, without benches. It is critical to let the medium dry between waterings, or else various diseases are likely. Some nurseries in the tropics grow stock plants outside, serving as windbreaks, as well as propagative sources. Cutting farms in the Caribbean and Latin America frequently grow aralias in the open in higher altitudes or under saran or natural shade at lower altitudes.

It is safest not to let the temperature go below 55F (12C), though some varieties, such as the Balfour types, can tolerate down around 40F (4C). Abrupt cold snaps tend to bother Ming and Parsley aralias. Most varieties are tolerant of humidity, though under high humidity, pressure from *Alternaria* increases. Most tropical production is done in shade or plastic greenhouses. Ventilation is important in preventing foliar disease problems.

NUTRITION

Most polysciases and aralias are moderately heavy feeders, with research indicating the maximum at around 1,500 pounds of nitrogen per acre per year (1,411 kg per ha). N-P_2O_5-K_2O ratios are generally 3-1-2 or 2-1-2. Constant liquid feed at such a ratio with 200 ppm nitrogen is frequently applied for smaller containers. For larger containers growers frequently use granular or slow-release fertilizers of a similar ratio, often including trace elements and sometimes magnesium.

Aralias in general don't have any particularly outstanding nutrient requirements. They are modest users of magnesium and trace elements, though sometimes additional magnesium helps highlight contrast in variegated varieties. *Polyscias* appears to be somewhat sensitive to ammonia, and leaf drop can occur if soluble salts are excessive.

DISEASES

Polysciases and aralias are susceptible to several common diseases if conditions warrant. They can all best be controlled with the right cultural conditions, including pathogen-free media. *Pythium splendens* is common under wet conditions and when growing aralias in moisture-retentive media. The plants wilt even when moist and show discolored roots. When you pull on a discolored root, the cortex or shell of the root comes off, leaving you a tan-colored, stringlike interior root part. This is not a foolproof diagnostic tool for *Pythium*, but it can be helpful. Drying the plants out between waterings can help, as do drenches of Subdue (metalaxyl) or Banrot (thiophanate methyl plus etridiazole), or sprays or drenches of Aliette (fosetyl-aluminum).

Rhizoctonia, another common cause of root rot, can also cause leaf spot and severe foliar blight under hot, wet conditions. *Rhizoctonia solani* is the most common type, and its symptoms are similar to those of *Pythium* when causing a root rot, though you may see reddish brown threads growing on the soil surface or around debris. The blight begins as a brown leaf spot, frequently with a yellow halo, and upper plant parts can turn dark brown very suddenly. Good potting media and moisture control are the best cultural controls. A soil drench with either Banrot (thiophanate methyl plus etridiazole) or Cleary's 3336 (thiophanate methyl) are somewhat helpful for root disease problems. Sprays of Zyban (mancozeb plus thiophanate methyl) are useful for foliar blight, as is Daconil (chlorothalonil).

Aralias are attacked by two types of *Alternaria*, *A. panax* and *A. araliae*. *A. panax* can be a little more aggressive. Both cause tiny, tan-colored leaf spots, usually circular, sometimes with a chlorotic halo. The spots can be reddish on variegated varieties, such as the Balfour types. *Alternaria* can cause foliar yellowing and occasionally severe defoliation in Ming and Parsley aralias. Keeping overhead irrigation and rainfall off of the foliage is very helpful. Keep your nutrient levels up, as the disease tends to attack hungry plants more aggressively. *Alternaria* sporulates better under high light, so when growing in bright shadehouses exposed to rainfall, *Alternaria* can be a problem. Sprays of Dithane (mancozeb) or Daconil (chlorothalonil) may be helpful, although there is one report of phytotoxicity of Daconil 2787 flowable (chlorothalonil) on Parsley aralia.

Anthracnose is another common leaf spot disease. It makes a larger spot than *Alternaria*, up to 1 inch (2.5 cm) across. Fruiting bodies of this fungus, which look something like black pepper granules, are frequently associated with the spots, which tend to be located near leaf margins and wound sites.

Keep the foliage dry. Sprays of Daconil (chlorothalonil), Dithane (macozeb), or Cleary's 3336 (thiophanate methyl) are helpful.

A bacterial leaf spot caused by *Xanthomonas campestris* pv. *hederae* is occasionally seen on aralias, especially variegated types. It is more visible on the underside of the leaf, appearing as a tiny, corky lesion, similar to that with edema (see "Disorders"). Bacterial lesions are no more than about one-sixteenth of an inch (1 mm) long and are dry in appearance. Keeping foliage dry is helpful, as are sprays of copper fungicides plus Dithane (mancozeb). Keeping reasonably good nutrition also helps reduce pressure from bacterial spots.

INSECT AND MITE PESTS

Two-spotted mites are the major pests in most parts of the world. Under hot, dry conditions the mites can proliferate rapidly, and due to the nature of aralia foliage, they can become quite severe before they are even noticed by the grower. Symptoms include foliar yellowing, speckling, and leaf drop. Use of miticides, including Avid (abamectin), Kelthane (dicofol), Pentac (dienochlor), is common on aralias. Good spray coverage is difficult on many aralia varieties, so a wetting agent can be helpful. Try to have enough spray pressure to turn the many small leaves over in order to achieve good coverage of the undersides.

Mealybugs are sometimes observed, typically seen as a cottony, segmented insect secreting honeydew and perhaps generating sooty mold. Sprays of Dycarb (bendiocarb), Dursban (chlorpyriphos), and to a lesser extent Orthene (acephate) are effective. Several sprays may be needed for good control. Scale insects at times create yellow blotches on the foliage, and several types of scales are reported. Control measures are similar to those for mealybug. Aphids can be troublesome, especially in spring. Sprays of Orthene or Dursban should control them.

 An expansion in a nursery is immediately followed by a temporary reduction in plant quality.

DISORDERS

Edema is a rather common disorder of many aralia varieties, especially the more succulent varieties with thicker leaves. The Ming and Parsley types tend not to have edema problems, but even with them edema is not unheard of.

Edema is a physiological disorder where moisture fluctuations temporarily cause too much moisture within the plant. The excess moisture exudes through the intercellular spaces, and an unsightly brown, tan, or reddish series of blisters can develop on the undersides of the leaves, becoming visible on the top as well. This especially happens in shadehouse or outdoor conditions, where moisture fluctuations are more prominent. There is really no cure for edema other than keeping moisture supplies more constant.

A second common disorder involves phytotoxicity from the insecticide-nematicide Vydate (oxamyl liquid). Vydate sprays on plants in this family tend to cause a witches'-broom or strap-leaf symptom, numerous small leaves originating from a single node on the stem. The number of shoots emerging from the stem may be normal, but the leaves may be strapped or distorted. Vydate is no longer registered on ornamentals in the United States, so the problem is no longer encountered here.

Severe leaf drop can be caused by *Alternaria*, elevated soluble salts, or moisture fluctuations.

TRICKS

The primary secrets to growing good polysciases and aralias are to have a potting medium that is not excessively high in moisture-holding capacity and to maintain a steady, even moisture supply. This of course is difficult in outdoor growing or under shadehouses. Keeping the light reasonably high is also helpful, although that tends to make the potential for *Alternaria* greater.

Most aralias are sensitive to ethylene exposure. Plants will drop lower foliage severely, especially when plants are placed close together with poor air movement. Waterlogged soil tends to generate ethylene, and if ventilation is poor or plants are not spaced, premature maturation of foliage and leaf drop occur.

INTERIOR CARE

Aralias in the interior like moisture and are sensitive to drying out, but be extremely careful to avoid excess watering. It is usually best not to apply too much water with any single irrigation. Leaf drop from irregular soil moisture is usually the major aralia problem in interiors.

When humidity is low the plants tend to get spider mites, which can become quite severe before they are noticed. The plant gets by with a minimum of 75 f.c. (0.8 klux), but they generally do better with significantly more than that.

REFERENCES

Bailey, L.H., and E.Z. Bailey. 1976. *Hortus Third*, 98. New York: Macmillan.

Burch, D.G., and T.K. Broschat. 1983. Aralias in Florida Horticulture. *Proc. Fla. State Hort. Soc.* 96: 161-164.

Henley, R.W., L.S. Osborne, and A.R. Chase. 1986. Polycias. *Foliage Digest* (March): 6-8.

ARAUCARIA

Norfolk Island Pine

HABITAT

The Norfolk Island pine, *Araucaria heterophylla,* is not a pine at all, but a member of a small, separate family, the Araucariaceae. The plant gets its common name from its native habitat, Norfolk Island in the Pacific. The araucarias in general are an important timber resource in the Southern Hemisphere. The Norfolk Island pine is said to reach over 200 feet (61 m),

Fig. 5 A 10-inch multiple araucaria in Homestead, Florida. (Courtesy of Ed Clay)

though in landscapes in California, Florida, and Hawaii, I have rarely seen it over 100 feet (30.5 m).

Norfolk Island pine is one of the few foliage plants that is a gymnosperm, a cone-bearing plant, in this case having both male and female plants. The family has been around for about 170 million years, and representatives can be found on most continents. There are about 15 species, with *A. heterophylla* (not *A. excelsa*) being the important species in the foliage trade. Some plants in the trade may actually be *Araucaria columnaris,* though, since the juvenile stages are very similar. The tree frequently grows near the coast, so it is one of relatively few salt-tolerant foliage plants.

Uses

Specimens of Norfolk Island pine are usually produced in containers ranging from 4 to 10 inches (10 to 25 cm), rarely larger. The plant is attractive as a single-stem item, though three plants per pot is also common, generally with the seedlings planted very close together in the container. Three plants per pot yields a somewhat fuller, wider, and more bushy appearance, whereas single-trunk specimens are more columnar. When well grown, Norfolk Island pine makes an attractive small potted Christmas tree, and some growers encourage this by decorating with red ribbons, ornaments, and foil. Norfolk Island pines are sold as foliage items throughout the year, but peak sales are usually in November in many markets.

Varieties

There are no named varieties of *Araucaria heterophylla,* to my knowledge. *A. columnaris* does show up in the trade occasionally as *A. heterophylla.* When mature, *A. columnaris* has darker foliage and closer tiers of branches. *A. heterophylla* is generally preferred as a landscape plant in USDA Zone 10.

Propagation

Araucarias can be grown in the summer months from cuttings, generally 2 to 8 inches (5 to 20 cm) in length, with the aid of 1,500 ppm indolebutyric acid (IBA), though I have never seen them done that way commercially. Most commercial propagation in the United States is from seed, frequently from Norfolk Island, Hawaii, or Central America. The seeds are harvested from female cones (strobili), which look something like large, green pine cones. The seeds are flat, light brown, and roughly 1.5 inches (4 cm) long. They are said not to store well, and they will germinate within themselves even if not planted. Depending on source, seed availability is from July to September in northern climates, January to March in southern climates. Seeds are generally broadcast or planted upright in flats or beds of moist soil.

 When a potted plant is growing outdoors in the Northern Hemisphere, root growth is usually best on the north side of the pot.

Seedlings generally come up in two to three weeks, and the emerged seedling quickly sheds the cotyledons, showing the first true leaf. Five to 10% albinism is common, and albino seedlings are generally discarded. Because of the rapid taproot development, seedlings need to be stepped up into pots rather quickly, or else the extensive taproot will become distorted or injured during transplanting, resulting in poor plants later. Seedlings should not be planted deeply, as Norfolk pine does not have the ability to generate adventitious roots. Seedlings planted deeply tend to develop stem rot.

CULTURE

Araucarias grow naturally in full sun and may be produced in nurseries that way, though the color tends to be rather pale. Pines can be grown somewhat faster in 30 to 63% shade. This is approximately equivalent to 4,000 to 8,000 f.c. (33 to 86 klux). The plants tend to become stretched and unattractive when grown in lower light levels; one technique is to produce them in brighter light early, then finish them briefly under extra shade to induce good color.

Peat-and-sand mixtures are common potting media, with the sand providing extra bulk density for growing larger specimens outdoors. This is important to prevent plants from being blown over in the wind and becoming damaged, a frequent occurrence. Optimum medium pH appears to be around 5.5, but the plants seem to grow well even with pH about 4.5. Norfolk Island pines are quite heat tolerant but somewhat susceptible to cold. Minimal cold injury can begin around 40F (4.5C), with significant damage to growing tips and needles as temperatures drop below freezing.

NUTRITION

Norfolk Island pines grow rather slowly when young, with growth rates increasing steadily with age. A 3-1-2 ratio of $N\text{-}P_2O_5\text{-}K_2O$ is common. Liquid or slow-release, coated fertilizers are normally applied when plants are young, as the plants have difficulty handling high rates of granular fertilizers. Once plants are more mature, though, they handle granular fertilizers well. Feed rates are also generally increased with increasing light levels.

Araucarias tend to have rather weak root systems when they are young, and extra phosphorus in the fertilizer program seems to help the root systems. Being somewhat acid loving, they develop iron deficiency at times, which appears as yellowing in the leaf tips, especially when medium pH rises above 7.0. Copper deficiency is also observed at times when growing in organic potting media. A distortion or crippling of the growing tip occurs, and the leaves appear to have trouble forming properly. One or two sprays of

TABLE 3 **Leaf analysis rating standards for *Araucaria heterophylla***

Nutrient (%)	Very low	Low	Medium	High	Very high
Nitrogen	<1.2	1.2–1.4	1.5–2.8	2.9–3.5	>3.5
Sulfur	<0.10	0.10–0.14	0.15–0.25	0.26–0.35	>0.35
Phosphorus	<0.15	0.15–0.19	0.20–0.30	0.31–0.45	>0.45
Potassium	<1.25	1.25–1.49	1.50–2.50	2.51–3.50	>3.50
Magnesium	<0.15	0.15–0.19	0.20–0.50	0.51–0.75	>0.75
Calcium	<0.50	0.50–0.69	0.70–1.50	1.51–2.25	>2.25
Sodium			<0.11	0.11–0.20	>0.20
(ppm)					
Iron	<40	40–49	50–300	301–500	>500
Aluminum			<101	101–250	>250
Manganese	<20	20–29	30–250	251–500	>500
Boron	<10	10–14	15–40	41–65	>65
Copper	<4	4–5	6–50	51–75	>75
Zinc	<12	12–19	20–200	201–300	>300

Sources: Institute of Food and Agricultural Sciences, Apopka, Florida; Dr. Benjamin Wolf, Fort Lauderdale, Florida.
Notes: Common name: Norfolk Island pine. Sample of most recent fully mature leaves, no petiole.

a copper-based fungicide will generally correct this situation, with the copper acting in this case as a nutritional supplement rather than as a fungicide.

DISEASES

By far the most troublesome disease of araucarias is *Colletotrichum* needle necrosis. The leaf tips turn dark brown to black, with tiny, black specks, the fruiting bodies of the fungus, observed on the necrotic tissue. The disease is more common in wet or windy weather, especially when the needles of plants rub together back and forth during storms. Sprays with Dithane (mancozeb) are effective, though they need to be applied regularly. Daconil (chlorothalonil) is also effective, but it is not labeled for Norfolk Island pine in the United States.

Pythium causes discolored, mushy root rot symptoms when plants are kept wet. Pines normally have only a few white roots at any given time, with the rest being a medium brown color. If they turn mushy, they may be infected with *Pythium*. Dry the plants down, and either spray with Aliette (fosetyl-aluminum) or drench with Subdue (metalaxyl).

Norfolk Island pines can also get *Cylindrocladium*, though not the same type that attacks *Spathiphyllum*. Applications of Terraguard (triflumizole) or Phyton 27 (PCAF) are helpful. Numerous other diseases are reported, but they are generally of minor consequence.

INSECT AND MITE PESTS

Fortunately, insect problems are not extremely common with this plant. Growers occasionally encounter mealybugs, which are usually controlled with multiple applications of either Dycarb (bendiocarb) or Orthene (acephate). There is a scale insect, *Oceanaspidiotus araucariae,* known to have attacked Norfolk Island pines in Hawaii, Florida, Puerto Rico, and elsewhere. Fortunately, the scale is not common, and controls are similar to those for mealybug. Thrips injury is also reported, but that is also an infrequent problem.

Microscopic mite injury is more common, observed at times when plants are under hot, dry conditions. The symptoms show up as distortion or chlorosis in the needles and in the growing tip of the apical shoot. Sprays of Kelthane (dicofol) or Thiodan (endosulfan) are helpful, especially with a surfactant or wetting agent. Avid (abamectin) also seems to help.

DISORDERS

Araucarias experience relatively few disorders during production, but when problems do happen, they can be rather difficult to diagnose. The effect of copper deficiency distorting the growing tip and affecting leaf development was discussed in the "Nutrition" section. High temperatures, 95 to 105F (35 to 41C), in conjunction with drying winds will cause damage to the growing tip and leaf tips. The tips of the leaves will appear to be burned off, and the leaves will frequently branch and begin growing again. To clarify, this is the tip of the leaf becoming necrotic, not the needle as with *Colletotrichum*. Shoot tip abortion can also occur when temperatures drop below freezing.

Leaves of Norfolk Island pine droop substantially when grown under excessively low light. Taproot distortion and curved or crooked trunks generally result when seedlings are kept in flats too long prior to transplanting. Careful staking can help somewhat when this occurs. Also, windstorms can cause significant physical damage and losses.

TRICKS

The Norfolk Island pine has a relatively weak root system, but the root system and the growth rate can be improved with regular foliar applications of Aliette (fosetyl-aluminum) approximately every six weeks. The plant also responds well in color and growth to regular sprays of soluble 20-20-20 fertilizer at about 2 pounds per 100 gallons (1 kg per 400 l).

Because they are produced from seed, Norfolk pine crops are genetically variable. Therefore, height differences of up to one-third are frequently encountered in a crop. The regular 20-20-20 sprays help the small plants somewhat more than the larger plants, and this helps make the crop more uniform. Araucarias grow slowly and steadily but at an increasing rate, so keeping fertility rates uniform is important for achieving high-quality growth and avoiding fluctuations in internode length.

Heavy wire stakes can be very helpful not only in straightening the trunk but in keeping the plants from blowing over. The wire stake is inserted through the medium and through the bottom of the pot, and into the ground about 6 inches (15 cm). The plants can then be tied to the stake, helping to straighten trunks and avoid damage from tipping over in the wind. It is also helpful to increase light levels after the plants are about 6 inches tall.

Araucarias can also be grown successfully when irrigation water is rather high in sodium and chloride.

INTERIOR CARE

The Norfolk Island pine is one of only a few conifers that will tolerate low light conditions. Interior light levels greater than 200 f.c. (2.2 klux) are best, though the plant can tolerate 100 f.c. (1.1 klux) for a period of time. It generally does not drop leaves when stressed. Fertilization should typically be avoided in interior situations, or else weak, stretched growth will occur, which is unsightly and will contrast with the compact nursery-grown leaves. After rough shipping conditions plants may need to be staked or restaked to help maintain an upright manner. Try to let the medium dry a bit between waterings. Because of the weak root system, repotting to larger containers is rarely necessary and usually undesirable.

REFERENCES

Bailey, L.H., and E.Z. Bailey. 1976. *Hortus Third*, 98. New York: Macmillan.

Hamon, A.B. 1988. *Oceanaspidiotus araucariae* (Adachi and Fullaway) (Homoptera: Coccoidea:Diaspididae). *Foliage Digest* (September): 8.

Neel, P.L. 1976. Araucarias. *Nurserymen's Digest* 10 (10): 118-120.

Osborne, L.S., C.A. Conover, and A.R. Chase. 1985. *Norfolk Island Pine*. AREC-A Foliage Plant Research Note RH-1985-H: 1-5. Apopka: University of Florida IFAS Agricultural Research and Education Center.

ARECA PALM

ASPARAGUS

HABITAT

The genus *Asparagus* is rather large, encompassing about 300 species of perennials. They range from herbs, vines, and shrubs to the edible vegetable *A. officinalis*. Most of the important cultivars are native to South Africa, which has very distinct wet and dry seasons. In their native habitat, ornamental asparaguses are found in areas of higher humidity and lower altitudes and where direct sunlight is reduced.

Ornamental asparagus cultivars are frequently called ferns, though they are not ferns at all, but members of the lily family. Ferns do not make flowers or seeds. Asparagus cultivars generally have an inconspicuous white flower and a red berry containing one seed.

Though they are foliage plants, asparaguses do not have true leaves. What appear to be leaves are actually modified, leaflike branches called cladophylls. For the sake of simplicity, I will refer here to the cladophylls as "leaves."

Fig. 6 *Asparagus sprengerii* in a hanging basket, Panama. (Photo by the author)

USES

As foliage plants, asparagus varieties are typically produced in 4- to 6-inch (10- to 15-cm) pots and in hanging baskets ranging from 6 to 10 inches (25 cm). The arching, trailing habit of several asparagus varieties makes them especially suited to hanging basket production. Some asparagus varieties—specifically, *A. setaceus*, commonly known as *A. plumosus*; *A. virgatus*, tree fern; and *A. medeoloides*—are also grown for cut foliage. Asparaguses are also frequently used as ground covers and for border plantings in tropical landscapes. They are hardy and durable as both houseplants and garden specimens.

VARIETIES

Most of the important varieties are cultivars of *A. densiflorus*, which is native to South Africa. The best known variety is *A. densiflorus* cv. Sprengeri, which is an arching, drooping, loosely branched plant with mint green foliage, small, white flowers, and red berries. *A. sprengeri* is somewhat spiny and has tuberous roots which help it deal with extended dry conditions. *A. densiflorus* cv. Myers (frequently Myersii) is a more stiff and upright variety; it is sometimes called foxtail asparagus or foxtail fern. The asparagus fern, known in the trade as *A. plumosus*, is actually *A. setaceus*. This variety is delicate and fernlike in appearance, with flat sprays of deep green foliage. *A. virgatus* is grown for cut foliage under the name of "tree fern" in Florida, Costa Rica, and elsewhere. It is somewhat more cold tolerant than other cultivated varieties. Finally, the florist's green known as smilax is *A. medeoloides*.

PROPAGATION

Commercial ornamental asparagus is propagated by seed. *A. sprengeri*, *A. myers*, and *A. setaceus* seed are readily available and usually easy to grow. Eight to 10 seeds are planted per pot or per liner, with germination typically running 80 to 90%. The seedlings come up in three to six weeks in warm soils (68 to 86F, 20 to 30C). A minimum of six seedlings per pot is usually required for adequate fullness in a production container or basket. Liners with fewer seedlings are generally combined in the pot to have a plant with adequate fullness. Clumps are also divided for propagation but not on a commercial basis.

CULTURE

Asparagus cultivars are generally of easy culture as long as a few basic guidelines are followed. They require reasonably bright light, in the range of 2,500 to 4,500 f.c. (27 to 48.6 klux). They tolerate less light but will grow more slowly. *A. setaceus*, the asparagus fern, is usually grown darker, more toward 80% shade, or 2,000 to 2,500 f.c. (21.5 to 27 klux).

Growth is slow below 55F (13C), but damage does not usually occur until temperatures drop below 45F (7.5C). *A. virgatus* is frequently irrigated for cold protection when temperatures approach freezing. Asparagus plants can be grown in a somewhat heavier, poorer potting mix than many foliage plants. Reed sedge peat (Florida peat) is often the principal medium ingredient, frequently with 30 to 40% pine bark or cypress chips. Asparagus plants tolerate a poorly drained medium better than many foliage plants, but only to a point. It is usually preferable to irrigate thoroughly, then let the soil dry down reasonably before resuming irrigation. This watering regime simulates what the plant frequently experiences in its native habitat.

NUTRITION

Asparaguses in general are not very particular in terms of nutrient requirements. N-P_2O_5-K_2O ratios of 3-1-2 are commonly used at the equivalent rate of

TABLE 4 Leaf analysis rating standards for *Asparagus densiflorus*

Nutrient (%)	Very low	Low	Medium	High	Very high
Nitrogen	<1.2	1.2–1.4	1.5–2.5	2.6–3.2	>3.2
Sulfur	<0.10	0.10–0.14	0.15–0.25	0.26–0.35	>0.35
Phosphorus	<0.20	0.20–0.29	0.30–0.50	0.51–0.75	>0.75
Potassium	<1.40	1.40–1.99	2.0–3.0	3.01–3.75	>3.75
Magnesium	<0.06	0.06–0.09	0.10–0.30	0.31–0.75	>0.75
Calcium	<0.06	0.06–0.09	0.10–0.30	0.31–0.75	>0.75
Sodium			<0.11	0.11–0.20	>0.20
(ppm)					
Iron	<40	40–49	50–300	301–500	>500
Aluminum			<101	101–250	>250
Manganese	<40	40–49	50–300	301–500	>500
Boron	<10	10–14	15–40	41–65	>65
Copper	<6	6–9	10–50	51–75	>75
Zinc	<16	16–24	25–200	201–300	>300

Sources: Institute of Food and Agricultural Sciences, Apopka, Florida; Dr. Benjamin Wolf, Fort Lauderdale, Florida.
Notes: Common variety names include Sprengeri, Myersii. Sample of most recent fully mature leaves, no petioles.

1,200 pounds of nitrogen per acre per year (134.8 kg N per 1,000 m² per yr). Nutrient deficiency symptoms are relatively rare, possibly due to the extensive root system and efficiency of uptake. When asparaguses get hungry, they turn pale, sometimes with some yellowing in the older foliage. Some necrosis and shedding of older leaves may also occur. Asparaguses do best with a steady nutrient supply, though they will tolerate low fertility for a period. Other than occasional iron deficiency symptoms (new leaves turning light yellow to white), trace element problems are rarely encountered.

DISEASES

Colletotrichum, commonly known as anthracnose, is the most common asparagus disease, resulting in yellowing and a light brown color in the middle to older leaves. It tends to occur when the foliage is moistened frequently. Sprays of Dithane (mancozeb) or Cleary's 3336 (thiophanate methyl) are frequently helpful. The fungus *Fusarium* can cause root rot problems, especially in conjunction with nematode injury. The roots become rather withered in appearance and turn a tan to light brown, occasionally darker. Soil drenches with Cleary's 3336 can be helpful. *Pythium* is also reported as a root rot fungus on many asparagus varieties, but it is not very common.

Asparagus plants are susceptible to lesion and root knot nematodes (*Pratylenchus* and *Meloidogyne*). The tree fern, *A. virgatus,* appears to be more susceptible than other varieties but not excessively so. Nematicides are frequently applied when growing asparagus plants for cut foliage.

 I have never seen magnesium toxicity in any plant.

INSECT AND MITE PESTS

Fortunately, insect pests are not frequently encountered on asparaguses. *A. setaceus* is fairly susceptible to scale, which is usually easily controlled by two or three weekly sprays of Cygon (dimethoate). Aphids can proliferate when new growth is plentiful. Most common insecticides are effective in that situation. Thrips cause a whitish, rasping injury of the foliage, which can be severe at times. Sprays of Cygon (dimethoate), Mavrik (tau-Fluvalinate), or any of several other common insecticides are helpful.

European red mites and two-spotted mites can attack asparagus plants, especially during warm and dry weather. Miticides, such as Pentac (dienochlor) and Avid (abamectin), and insecticidal soaps are commonly used for mite control.

DISORDERS

Relatively few disorders of asparagus plants are encountered. When humidity drops quite low, especially in interior conditions, leaf drop can occur. Asparaguses are very sensitive to fluoride in the atmosphere and will defoliate if exposed to fluoride air pollution. Extreme drying can also cause defoliation, as can heavy mite or thrips injury. Branch dieback is usually related to irregular watering or fungal root rot.

TRICKS

Asparagus varieties generally do not give the grower a great deal of trouble, as long as they receive good light levels and regular fertilization and watering. When *A. sprengeri* becomes jostled during shipping or when plants are very old, the branches tend to tangle somewhat, giving an unattractive appearance. A common garden cultivator or claw is very useful for combing the branches to give *A. sprengeri* a neater appearance.

INTERIOR CARE

Reasonably bright light, 150 to 250 f.c. (1.6 to 2.7 klux), is best for asparaguses. The minimum light level they can tolerate for extended periods is 100 f.c. (1.1 klux). As in a growing situation, irrigate reasonably well and let the soil dry out. Maintain temperatures above 50F (10C). If the bottom of the pot has a saucer, it is best to tip the pot to drain the saucer after watering, or root rot can develop. Asparagus plants can last almost indefinitely in interior situations with good care. With their aggressive root systems, they do tend to get pot bound faster than many foliage plants. When that occurs, it is helpful to divide the plants or place them in larger containers. Asparagus plants will survive when extremely pot bound, though the foliage canopy tends to become rather thin.

REFERENCES

Bailey, L.H., and E.Z. Bailey. 1976. *Hortus Third*, 118-119. New York: Macmillan.

Platt, W.J. 1952. *Asparagus Fern Culture*, 1-11. Agricultural Extension Service Bulletin 153. Gainesville: University of Florida Agricultural Extension Service.

Stamps, R.H. 1987. Herbicides and Treefern. *Nursery Digest* (November): 70-71.

BAMBOO

(See Other Foliage: Bambusa*)*

BAMBOO PALM

(See Chamaedorea*)*

BIRD-OF-PARADISE

(See Other Foliage: Strelitzia*)*

BROMELIADS

HABITAT

Bromeliads, named after the Swedish botanist M. Bromel, have been popular in European greenhouses for the last 150 to 200 years. Bromeliads have been popular in the United States really only since the mid 1970s, though a few producers were doing them on a small scale earlier. In the same family as the pineapple, bromeliads may be either epiphytic, terrestrial, or in some cases both. They use their roots primarily for anchoring, as opposed to nutrient absorption, and their cultural requirements are somewhat more similar to cacti and succulents than to other foliage plants.

Fig. 7 *Guzmania* and *Aechmea*, two popular bromeliads, in bloom. (Photo by the author)

The approximately 45 bromeliad genera contain about 2,000 species. They are mostly native to tropical and subtropical regions of North, Central, and South America. The epiphytic types frequently grow in trees along riverbanks. Terrestrial types are found in a wide variety of habitats. *Aechmea* may be found in the wild growing in trees, on rocks, and in the ground from Mexico to Argentina. *Billbergia* was among the first popular cultivated bromeliads in the United States. It is usually epiphytic, though it will grow in coarse humic and mulch types of environments, such as forest litter. *Guzmania* prefers the shady, damp locations within rain forests, primarily in Colombia and Ecuador. It is less heat tolerant than other bromeliads and prefers lower light conditions. The better *Neoregelia* cultivars are found in southeastern Brazil and the Amazon region. Their multicolored foliage does best in filtered light between 50 and 70% shade and without excess fertility. *Nidularium* is found both in trees and on the ground in the humid forests of eastern Brazil, so it therefore naturally prefers lower light levels and moist, humid conditions.

The durable, gray-colored *Tillandsia* is found growing epiphytically from Florida to Argentina in many different types of habitats. It generally tolerates a wide variety of conditions, but not poorly drained soils or wet conditions. The habitat of *Vriesea* ranges from Mexico to Argentina, with the best cultivars coming from Brazil. It also dislikes excessively wet conditions, and its sometimes-variegated foliage does better under lower light levels.

USES

If you are looking for exotic, hardy, low-maintenance, attractive interior foliage plants which require little water, bromeliads are the ticket. Their long-lasting flowers are often stunningly beautiful, and most cultivars have attractive foliage even when the plants are not flowering. There is a good, consistent, year-round demand for bromeliads, though peak sales are usually the holidays from Christmas to Mother's Day.

Pot sizes ranging from 4 to 6 inches (10 to 15 cm) are the most common. Epiphytic types are also grown on plaques, on pieces of driftwood, and in other imaginative presentations. I have seen *Tillandsia* grown for refrigerator magnets. The larger bromeliads are sometimes used as living vases for cut flowers. Planting bromeliads in nature gardens with rocks and moss enables the grower to use imagination and design talents. Bromeliads are popular in mass plantings for interior settings as well as for small, flowering accent plants for tables and desks. Clustered bromeliads are very striking when planted appropriately in tropical landscapes.

VARIETIES

Probably the best known bromeliad variety is *Aechmea fasciata*, the silver vase. Its striking silver-and-white foliage and large, contrasting, pink bloom is well known to consumers. Friederike is a popular *Aechmea* with darker foliage and a large, red flower. *Guzmania* Cherry is an attractive, somewhat smaller plant with brilliant, reddish flowers. Other popular *Guzmania* varieties include Claret, Grapeade, Luna, and Ultra. *Neoregelia* is somewhat more low growing, frequently with variegated foliage and red flowers. Popular varieties include Meyendorffii, Tricolor, and Flandria. The small, branching Fireball is one of my favorites.

The bird's-nest bromeliads, represented by the genus *Nidularium*, frequently have striped or spotted foliage, with rather small but strikingly unusual flowers. *Tillandsia* is diminutive and somewhat less spectacular than other types, but it is very hardy and does well on plaques and pieces of wood. *Vriesea*, commonly called the Flaming Sword or Painted Feather bromeliad, has multicolored or striped foliage and frequently large, flat flower spikes. Cultivars of Vriesea splendens are very popular and include Poelmanii, Splendide, and Splenriet. Numerous other attractive, useful bromeliads exist in the trade, too many to name here. New varieties are also coming all the time. I've seen one Guzmania with a yellow flower that lasts nine months!

PROPAGATION

Bromeliads may be propagated from seeds, pups (suckers), or tissue culture. Some growers maintain stock plants of various varieties and harvest the suckers for commercial production. For larger quantities, however, it is more common to purchase either rooted or bare-root liners from either tissue culture or seed. There are some good U.S. producers of bromeliad liners, though many growers still prefer to purchase plantlets in quantity from Europe, especially Holland. Boxes of liners should be promptly opened and planted. Follow the supplier's planting directions precisely.

CULTURE

Bromeliad culture varies quite a bit by variety. They tend to prefer a lower pH range in the medium. Selecting a good potting medium is a critical step for a good bromeliad grower. Don't try to grow bromeliads in a basic foliage mix, or in a peaty mix. Instead, use a coarse, chunky mix that falls apart when you dump it out of the pot. It is important to have wood sources, such as cypress, redwood, or bark, in the mix, along with perlite, osmunda, and other coarse materials with minimal moisture-holding capacity. The cups (whorls) of bromeliads are maintained with water in them in greenhouses, though less so in the interior environment.

Bromeliads are rather slow, sometimes taking 18 months or even up to three years to produce salable plants. However, they are minimal users of greenhouse space in their early stages, so the slow turnaround time can be compensated for by high plant density.

The best temperatures for bromeliads are 55 to 85F (12.8 to 29.4C). Minimum night temperatures of around 50F (10C) are best for many varieties. Frequently, plants exposed to lower temperatures than that will not flower. *Neoregelia* is often grown in 50 to 70% shade, with *Aechmea* sometimes a little brighter. *Guzmania* prefers cool, shady conditions with good air movement and humidity. The pH is best maintained around 5.0 for most varieties. Flowering is usually induced on year-and-a-half to two-year-old plants by spraying Florel (ethrel) at 2,500 ppm. This is equivalent to 8 ounces per gallon (65 ml per l). Flowers will follow in about two months.

NUTRITION

Because of their epiphytic nature, many bromeliads are able to absorb nutrients through the peltate scales in their leaves. In their native habitats the plants feed on decomposing plant and insect debris that falls into their cups. (Tree frogs frequently lay their eggs in bromeliad cups high up in large trees in the tropical rain forests.)

For the grower this means liquid feed is useful for bromeliads, but many growers have also gone to incorporating or top-dressing coated, long-term, slow-release fertilizers, supplemented with liquid feed. Ratios of 1-1-1 are popular, though many growers tweak the fertilizer program with extra phosphate or potash, depending on their experience with different varieties. Deficiency symptoms are not very common, but *Aechmea* has a fairly high magnesium requirement, and many cultivars respond to extra potassium. *Guzmania* is rather sensitive to excess boron, as are several other varieties. The Europeans often suggest that you discontinue liquid feed when bromeliads begin to bloom. In Florida, however, that has generally resulted in smaller, less spectacular flowers.

DISEASES

A wide variety of diseases are reported on bromeliads, but with decent culture most growers find the plants to be relatively pathogen free. *Pythium* root rot is common when plants are overwatered or if the mix is too absorbent. Dry the plants down, go to a less absorbent mix in the future, and drench with Subdue (metalaxyl). *Fusarium* and *Rhizoctonia* can also cause root rot under similar conditions to *Pythium.* Usually, a drench with one of the thiophanate methyl fungicides helps control it. *Helminthosporium* is a common fungal leaf spot on *Aechmea* and other varieties. Good moisture management and sprays of Dithane (mancozeb) are helpful.

Neoregelia is sometimes attacked by *Cephaleuros virescens,* an algal leaf spot. This is very difficult to control, though sprays with quaternary ammonium materials, such as Physan, may be helpful. When excessively moist, many bromeliads suffer from edema, where excessive moisture exudes from the older leaves, leaving leaf spot symptoms behind. Maintaining steady moisture levels and a consistent greenhouse environment avoids this. Problems with viruses and nematodes are extremely rare.

INSECT AND MITE PESTS

Mealybugs are the major insect problem for bromeliad producers, and they can be quite troublesome. They are generally introduced by plants brought into the greenhouse, so inspect all inbound shipments thoroughly. The insects may hide in the cups, between leaves in the whorl, and on roots. Malathion is frequently used to combat them, as is Cygon (dimethoate) on some varieties. Phytotoxicity information on bromeliad varieties is very limited, so you need to be careful and test unfamiliar sprays, or injury can result. Scale insects, including the stellate scale, are also occasionally encountered. Scale insects are usually found on the undersides of older leaves, where their feeding causes yellow blotches. Treatments are similar to those for mealybugs.

Spider mites occasionally cause problems, for example, the pineapple mite on *Aechmea*. The mites feed down in the cup, and as the leaf emerges, pale yellow clusters of spots are visible. The problem looks somewhat like a disease and is frequently mistaken for a fungal leaf spot. Sprays of Avid (abamectin) or Kelthane (Dicofol) are helpful, but again, check on phytotoxicity.

The bromeliad pod borer, *Epimorius tesluceellus,* is occasionally a severe problem, but usually only on *Tillandsia fasiculata*. Mosquitoes often breed in the cups of bromeliads grown in shadehouses. This generally doesn't go down well with nursery workers. Occasional sprays of Malathion over the top help keep the mosquitoes in check.

DISORDERS

Copper in any form is toxic to just about all bromeliads. Severe foliar necrosis may develop on any part of the plant, and the entire cup may die back. Avoid copper fungicides, copper wood preservatives, and trace element preparations with significant amounts of copper. The little bit of copper found in most soluble fertilizers will generally not cause a problem.

Bromeliads grown in excessively moist media frequently wobble in the pots due to inadequate rooting. Drying the plants down and changing the media usually corrects this. Many bromeliads can survive fairly well for a time without any roots at all, though the plants are unattractive and will fail to flower properly.

Aechmea fasciata develops necrotic spots on the foliage if shipped at low temperatures. Damage occurs at 50F (10C), but 68F (20C) is fine. *Neoregelia* Tricolor does better when shipped cool. If flower buds drop or fail to open, the cause is usually low humidity. Increasing humidity or misting the plants usually helps the blooms open. When exposed to cold temperatures or cold irrigation water, the foliage of many bromeliad varieties withers and quickly becomes unattractive.

TRICKS

I frequently tell growers that you may be in the foliage business, but what you should really be trying to grow is roots. Grow good roots, and most of the rest of your problems will take care of themselves. Don't depend on feeding bromeliads through the cups. Develop a good root system by decreasing the peat percentage in your mix and increasing the percentage of wood products. Drench at potting with Subdue (metalaxyl) and a high-phosphate starter fertilizer, such as 9-45-15 or 10-52-10, in order to stimulate good root growth. You will be amazed at the difference.

When tying bromeliads onto driftwood and plaques, you must use stainless steel or covered wire, because any copper in the wire may be toxic to the plants. Many bromeliad varieties do well in clay pots with straight cypress mulch as the media. Air movement is very important with bromeliads. If you think about it, many of them grow up in trees where air movement is relatively constant. Try to duplicate that in your greenhouse.

INTERIOR CARE

Except for a few varieties, bromeliads need relatively high humidity to do well. They are tough, durable interior plants if their minimum requirements are met. Don't try to keep the cups wet all the time in interior environments. It is best to water the medium when it is very dry, and splash only a little bit in the cups. Change the water in the cups periodically by turning the plant upside down and draining the water. Otherwise, fungi and algae may proliferate.

A minimum of 200 f.c. (2.2 klux) is needed for most varieties. You can fertilize with about one-third teaspoon per gallon of soluble 20-20-20, but only if the light levels are 750 f.c. (8 klux) or more. Don't fertilize in the dark. Better flowering and better longevity are achieved under higher interior light levels. The best temperature range is 55 to 85F (12.8 to 29.4C).

Bromeliads brought home in bud frequently have attractive flowers for three to six months indoors. After blooming, many varieties send out suckers, which can be easily transplanted.

Hose down walkways to increase greenhouse humidity.

REFERENCES

Anderson, R.G. 1984. The Color Alternative. *Interior Landscape Industry* (May): 36-41.

Fairchild, D. 1988. What Is a Bromeliad? Eighth World Bromeliad Conference. Miami: Bromeliad Society of South Florida.

Frank, J.H. 1991. Bromeliads and Mosquitoes. *Foliage Digest* (April): 7-8.

Heppner, J.B. 1992. *Bromeliad Pod Borer,* Epimorius testaceellus. Entomology Circular 351, FDACS: 1-2. Gainesville: Florida Department of Agriculture and Consumer Services, Division of Plant Industry.

Kent, M. 1992. Grower's Notebook. *Greenhouse Manager* (January): 12.

Klenn, J. 1984. Designing With Bromeliads. *Florist's Review* (July 12): 29-30.

Poole, R.T., R.W. Henley, and C.A. Conover. 1991. Necrosis of Bromeliads During Storage and Shipping. *Foliage Digest* 14 (1): 1-2.

CACTI

(See Succulents: Cacti)

CALADIUM

HABITAT

Most of the important *Caladium* cultivars are native to wet areas of the Amazon basin in Brazil. Curiously, the name of the genus is of East Indian origin. Caladiums are members of the family Araceae, as are several other foliage plants, such as philodendron, anthurium, syngonium. They thrive best in relatively acid, organic soils and are frequently found in the wild and in field culture growing at a soil pH of 4.5.

USES

Caladiums, which are fairly easy to grow, are produced in smaller

Fig. 8 Multiple caladiums in a bulb pan. (Photo by the author)

pots, typically 3, 4, and 6 inches (7.5, 10, and 15 cm). They are a quick-turn crop, usually being completed within five to eight weeks. The heart-shaped, multicolored foliage makes the plant very popular for Easter and Mother's Day pot sales. The market is year-round, though, with emphasis on spring and summer sales.

VARIETIES

The first hybrid caladiums were introduced to horticulture in 1867. As early as 1910 Dr. Nehrling of Orlando had as many as 2,000 varieties, many of which resulted from his own crosses. Today there are about 100 or so varieties in commercial culture, with about 20 varieties making up the main body of the market. Some prominent cultivars include Candidum (white with green veins), Frieda Hemple (predominantly pink), Fannie Munson (predominantly red), and the lance variety White Wing. Varieties which tolerate indoor conditions include Lord Derby, White Christmas, Fire Chief, Red Flash, Carolyn Whorton, Poecile Anglais, Sea Gull, Scarlett Beauty, and Aaron.

Commercial cultivars known as the fancy-leaved caladiums are normally hybrids of *Caladium × hortulanum*. Some are also *C. picturatum* cultivars, the modern lance- or strap-leaf varieties. These are shorter plants with smaller leaves that are more ruffled along the edges. Most caladium tuber (bulb) production today occurs in Highlands County, Florida, in the moist, organic soils around Lake Placid.

Propagation

Most caladiums are produced from tubers, commonly called bulbs, about 95% of which are field grown in Central Florida. They can be grown from seedlings, but plants from seed are slow to finish and are generally not produced. Propagation has been done from tissue culture from time to time, but cost factors and varietal instability have been limiting factors in tissue-culture caladiums.

The tubers are harvested from the field and stored for eight weeks. Tubers should not be exposed to temperatures below 60F (15C). When they arrive from the supplier the tubers should feel firm, rubbery, and somewhat sweaty. If they are spongy, it means they have been exposed to temperatures below 60F. The tubers should have been stored at 70F (21C) for eight weeks, and not less than six weeks.

Many growers prefer to gouge out or scoop the dominant eye of the tuber in order to get more shoots of uniform size. This results in a somewhat fuller plant, though with smaller leaves, and gives a slight delay in sprouting.

Culture

Being strictly tropical, caladiums should be grown under warm conditions, ideally 80 to 90F (26.5 to 30.2C) but a minimum of 70F (21C). Bottom heat is common in northern climates. Research has shown that increasing the temperature from 70 to 90F (21 to 32C) speeds the crop up only a little bit, so in more temperate climates, you can stay at the lower end of the temperature range. Minimum night temperatures should be 65F (18.3C). Plants suffer injury at 55F (12.8C), and most are killed when exposed to 35F (2C).

Light levels are usually between 60 and 80% shade, equivalent to 2,500 to 5,000 f.c. (6.9 to 53.8 klux). Color pattern of the varieties changes with light level, and you should experiment with different light levels to find the ideal range for your cultivars. Candidum is frequently grown at 1,000 f.c. (10.8 klux), whereas the red and pink varieties are produced at the brighter end of the range.

Crop times vary somewhat by planting date and by the number of tubers per pot, but most plants can be finished in five to eight weeks. The number of tubers varies with variety and cost factors and whether the tubers are scooped. A 6-inch (10-cm) pot is frequently planted with three to five #2 tubers. Smaller pots may have only one or two tubers, especially when scooped. Planting depth should be 1 to 1 1/2 inches (2.5 to 3.8 cm). Caladiums are tolerant of a wide range of potting media. Two parts peat to one part sand are used in some areas, as are mixtures of two parts peat, two parts bark, and one part sand. Shoot for a pH of 5.5 to 6.5.

NUTRITION

Unlike many foliage plants, caladiums prefer a 2-2-3 ratio of N-P_2O_5-K_2O or a 1-1-1 ratio. Granular 6-6-6 or slow release 14-14-14 are frequently used at about one teaspoon per 6-inch (5-cm) pot. High nitrogen applications are discouraged, as the plant tends to grow very weak and stretched, with poor color. Growers using liquid fertilizer frequently use a constant liquid feed of about 150 ppm nitrogen and 150 to 200 ppm K_2O.

The caladium is one of only a few foliage plants where nutritional factors and deficiency symptoms have been well researched. The plant has relatively high requirements for potassium, magnesium, calcium, and boron. See the "Disorders" sections for nutrient deficiency symptoms.

DISEASES

Like most other aroids, caladiums are susceptible to dasheen mosaic virus. Incidentally, tissue culture was originated as a method to obtain virus-free aroids, and it was then realized that tissue culture was also a useful means for propagating. Growers have done a good job in reducing or eliminating the virus from stock; therefore, dasheen is not a major problem today in caladiums.

Root knot nematodes are a significant problem for tuber producers, but soil fumigants and hot water dips (50C, or 122F, for 30 minutes) are effective against nematodes. *Fusarium solani* causes a common chalky tuber rot. A 30-minute soak in a thiophanate methyl fungicide (in the U.S.; Benomyl elsewhere) is helpful. In hot weather southern blight, caused by the fungus *Sclerotium rolfsii*, can kill plants. The fungus is characterized by white threads growing on soil and plant surfaces, with white to tan fruiting bodies (sclerotia) that are quite visible. Sprays or drenches with either Terraclor (PCNB) or the insecticide Dursban (chlorpyriphos) are effective. No, I am not kidding. Tubers will also rot when exposed to temperatures below 50F (10C).

INSECT AND MITE PESTS

Aphids and thrips are the most common insect problems on *Caladium*. Aphids of course feed on the new leaves and cause sucking injury and puckering of the foliage. Thrips frequently create rasping injury in the unfurling leaf, and damage is often worse on one side of the leaf than the other. Malathion is frequently sprayed to control these pests, as are Orthene (acephate) and Mavrik (tau-Fluvalinate). Mealybugs, which can attack tubers or potted plants, are generally controlled with Dycarb (bendiocarb) or Cygon (dimethoate).

Two-spotted mites can be a problem in warm weather, giving you the typical small, white specks on the foliage from their feeding injury. Sprays of Pentac (dienochlor), Avid (abamectin), or insecticial soaps are useful. Silverleaf whiteflies have become a production problem in recent years in many areas. The best control measures vary with the population of whiteflies, but useful materials include the neem extracts (such as Azatin), combinations of Orthene (acephate) and Talstar (bifenthrin), or Marathon (imidacloprid).

DISORDERS

Many caladium cultivars, especially the white ones, show foliar burn symptoms when drought-stressed or exposed to high light. Large, oval lesions form in the interior part of the leaf, starting out with a translucent appearance and usually ending up as a tan lesion or a hole in the leaf.

Nitrogen-deficient caladiums show chlorosis and reduced leaf size. When lacking in phosphorus, the plants are small and stunted. Potassium deficiency is characterized by marginal chlorosis and foliar lesions, especially in older leaves. Calcium-deficient plants have reddish brown spots on the underside of the leaf near the petiole. Caladiums lacking magnesium have chlorosis and holes in the older leaves, while iron-deficient plants have yellow new leaves, but with green veins. When lacking manganese the plants turn an odd yellow-green color, with small, speckled new leaves. Plants low in boron display brittle petioles, which tend to break halfway, leaving the leaves dangling.

TRICKS

Some growers like to plant caladium tubers upside down in order to get more green in the finished plants, though for most cultivars this practice is discouraged. Too many growers put all of their caladium crops in the same house under the same light levels. In reality, to grow excellent-quality caladiums, you need to experiment somewhat to find the best light level for your cultivar, as color pattern and intensity vary substantially. Truly well grown caladiums are produced at just the right light level.

When grown as bedding plants, the varieties that tolerate full sun generally have at least one-third green coloration in the foliage. Store tubers at high humidity, but don't ever refrigerate them. Surflan (oryzalin) is a very good preemergent herbicide for this crop.

INTERIOR CARE

Protection from cold drafts is critical for this truly tropical foliage plant. Don't keep caladiums near doors or windows which open frequently. For best color and growth, give them as much light as you can, without exposing them to direct sunlight. Try not to let the temperature drop below 65F (18.3C).

Caladiums frequently die back and rest for a while. This usually happens during winter, and you can withhold water for some time to let the tubers rest, then revive them in the spring with water and fertilizer. Caladiums can easily be damaged by cold irrigation water in northern climates, so let the water be at room temperature before irrigating.

Rainwater is better than groundwater because it is very pure and saturated with air.

REFERENCES

Black, R.J., and B. Tjia. 1979. *Caladiums for Florida.* Fla. Coop. Ext. Serv. Circular 469. Gainesville: University of Florida Institute of Food and Agricultural Sciences.

Gilreath, J.P., and B.K. Harbaugh. 1985. Chemical Weed Control in Caladiums. *Hortscience* 20 (6): 1056-1058.

Harbaugh, B.K., F.J. Marousky, and J.A. Otie. 1979. Warm Environment: Key to Forcing Caladiums. *Florist's Review* (January 11): 64-68.

Harbaugh, B.K., and B.O. Tjia. 1984. Size and Scoop Caladium Tubers. *Greenhouse Grower* (August): 82-83.

Harbaugh, B.K., and B.O. Tjia. 1987. Diagnosing Nutrient Disorders in Caladium. *Greenhouse Manager* (October): 109-114.

Knauss, J.F. 1975. Description and Control of Fusarium Tuber Rot of *Caladium. Plant Disease Reporter* 12 (December): 975-979.

Larson, R.A. 1984. Caladiums. *Florist's Review* (July 5): 33-34.

Sheehan, T.J. 1987. *Caladium Production in Florida.* IFAS Circular 128B. Gainesville: University of Florida Institute of Food and Agricultural Sciences.

Wilfret, G.J. 1984. Caladiums to Know and Grow. *Foliage Digest* 7 (7): 1-3.

CALATHEA

HABITAT

Among the most colorful of the tropical foliage plants are the calatheas, occasionally known as peacock plants or rattlesnake plants. About 100 species of *Calathea* exist. Generally native to tropical America, especially Brazil, calatheas grow best in their native habitat as understory plants in humid jungles.

The genus name comes from the Greek word for "basket." Calatheas have many leaf colors and variegation patterns. The plants belong to the family Marantaceae and are sometimes confused with marantas and ctenanthes.

USES

Calatheas are very popular in Europe due to their colorful qualities, even under low light conditions. Many flowering foliage plants do not produce color under extremely low light, but even without a flower calatheas are very striking and popular. Pot sizes range from 3 to 4 inches (7.5 to 10 cm) for smaller varieties, up to 10 inches (25 cm) and even larger for varieties such as *C. zebrina*, which can easily be 3 feet (0.9 m) tall. Dwarf varieties are useful in terrariums. Low humidity can sometimes become a limiting factor in interiors.

Fig. 9 *Calathea* production in a double-poly greenhouse. (Photo by the author)

VARIETIES

A number of popular calathea varieties exist. Some of the more common ones include *C. insignis,* the rattlesnake plant, whose narrow, leathery foliage is rather wildly colored, with green, oval spots and a maroon leaf underside. The peacock plant, *C. makoyana,* has a cream-colored upper surface with green lines, is purple underneath, and grows about 1 foot (30.5 cm) tall. *C. louisae,* whose native habitat is unknown, is a bushy variety with yellow markings and purple underneath.

Some favorite larger varieties include *C. zebrina* and the somewhat similar *C. warscewiczii.* These plants reach 3 feet (0.9 m) or more in height, with velvety, iridescent foliage displaying yellow-green veins. *C. vietchiana* is a tall, open plant with connecting half-moons of contrasting colors. *C. ornata* cv. Rosea-lineata and *C. roseopicta* are larger-leaved varieties for larger containers. The former has pink stripes with burgundy underneath, and the latter has a more round leaf whose dark green is highlighted by a red midrib.

A rather unique variety is *C. crocata* Eternal Flame. It sports numerous orange flowers, which are similar to those of *Neoregelia* bromeliads. Most calatheas have rather inconspicuous flowers, but not this one. Among popular dwarf varieties, *C. micans* has pointed leaves and grayish green color underneath, and *C. undulata* has leaves that are wider and purple underneath.

 Many growers incorrectly believe that subjecting plants to moisture stress increases root growth.

PROPAGATION

Calathea propagation may be from seed, tissue culture, or division, depending on the variety. Calatheas grow and sport leaves from an underground stem or rhizome. The cultivars of *C. picturata, C. vandenheckei,* and *C. argentea* come true from seed. Several of the popular cultivars are multiplied in tissue culture laboratories, a method becoming increasingly popular. Plants can also be grown from division, typically two to three plantlets potted together in a 4-inch (10-cm) container. Divisions are usually placed under occasional mist at 700 to 1,000 f.c. (7.5 to 10.8 klux). Temperatures for young plants and divisions should be kept moderate, preferably around 70F (21C) at night and 82F (28C) during the day.

CULTURE

Calatheas are not considered particularly easy to grow. To produce them successfully, you must be rather precise in managing your greenhouse environment and nutrition program. They become much easier to grow if you simply modify the growing conditions to their liking. Popular potting media include a 3:1 mixture of peat and sand, a 1:1 mixture of sphagnum peat and pine bark, or in the western United States a mixture of one-third peat, one-third perlite, and one-third redwood sawdust. The plant prefers to grow at lower pH, around 5.5, though many growers must increase pH to around 6.5 if they have fluoride in their irrigation water or fertilizer. Incorporating trace elements into the potting medium is a good idea for this plant.

Try to maintain humidity at 60% or above, and don't go below 65F (18.5C) or above 90F (32C). Calatheas are not tolerant of heat. Mature plants can tolerate temperatures in the low 50s F (around 12C). With a well-drained, somewhat forgiving potting medium, they respond well to regular watering. Light levels are typically 1,000 to 1,500 f.c. (10.5 to 16 klux), depending somewhat on cultivar.

NUTRITION

Some growers like to produce calatheas at a constant liquid feed of 300-70-250 ppm N-P_2O_5-K_2O. Others like to incorporate slow-release fertilizers, such as a 20-10-15 or similar N-P_2O_5-K_2O ratio. Eight grams of 19-6-12 Osmocote per 5-inch (12.5-cm) pot has been a successful rate. *C. louisae* and *C. warscewiczii* are somewhat heavier feeders.

Low fertility can cause loss of color and leaf spotting, while high-potassium fertilization can reduce color intensity. Florida research has shown that at least half of the nitrogen should come from urea or ammonia sources, perhaps because of their acidity. I suspect growers in more northern climates would want to grow with more nitrate nitrogen. Most of the cultivars are somewhat sensitive to soluble salts in the media.

DISEASES

When calatheas are kept too wet, *Pythium* root rot is rather common. Drenches with Truban (etridiazole) or Subdue (metalaxyl) can be helpful, as are sprays of Aliette (fosetyl-aluminum). Drying the plants out of course is

TABLE 5 Leaf analysis rating standards for *Calathea* spp.

Nutrient (%)	Very low	Low	Medium	High	Very high
Nitrogen	<2.0	2.0–2.4	2.5–4.0	4.1–5.0	>5.0
Sulfur	<0.13	0.13–0.19	0.20–0.40	0.41–0.60	>0.60
Phosphorus	<0.15	0.15–0.19	0.20–0.50	0.51–0.75	>0.75
Potassium	<2.00	2.00–2.45	2.50–4.50	4.55–5.50	>5.55
Magnesium	<0.20	0.20–0.24	0.25–0.60	0.61–1.00	>1.00
Calcium	<0.25	0.25–0.49	0.50–1.50	1.51–2.50	>2.50
Sodium			<0.11	0.11–0.20	>0.20
(ppm)					
Iron	<25	25–29	30–200	201–500	>500
Aluminum			<101	101–250	>250
Manganese	<21	21–29	30–200	201–500	>500
Boron	<12	12–17	18–50	51–75	>75
Copper	<4	4–5	6–200	201–500	>500
Zinc	<15	15–19	20–200	201–500	>500

Sources: Institute of Food and Agricultural Sciences, Apopka, Florida; Dr. Benjamin Wolf, Fort Lauderdale, Florida.
Note: Sample of most recent fully mature leaves, no petioles.

critical. *Botrytis* can sometimes attack in cooler climates. Ornalin (vinclozolin) or Daconil (chlorothalonil) sprays are used along with cultural controls.

The common *Helminthosporium* leaf spot, caused by *Drechslera setariae,* is characterized as large, tan leaf blotches about one-half inch (1.25 cm) in diameter, usually with a halo. The foliage needs to be wet for six hours for this fungus to get a foothold, so try to irrigate when drying conditions are good. Different *Calathea* varieties have different levels of susceptibility to this disease, though disease incidence is not affected by fertilizer rates. Sprays of Daconil (chlorothalonil) or Dithane (mancozeb) will help control *Drechslera*. A third common foliar disease is *Alternaria alternata,* a small leaf spot about 2 mm across, frequently water-soaked initially and without a halo. Keep the foliage dry and spray with Daconil or Dithane. Cucumber mosaic virus causes jagged, yellow patterns in the leaf, which are not terribly noticeable, and doesn't cause any other symptoms. Insects usually spread it. Don't propagate from plants showing viral symptoms.

Burrowing nematodes *(Radopholus)* can be very serious. They cause root rot symptoms and root lesions along with severe stunting. Control is difficult, so try not to allow these pests into your nursery. Mocap (ethoprop) and Oxamyl can help, though they are not registered in many areas, including the United States.

INSECT AND MITE PESTS

Calatheas are affected by the usual battery of occasional pests. Spider mites cause speckling on the foliage and are usually controlled with Avid (abamectin), Pentac (dienochlor), or Kelthane (dicofol). Mealybugs and scale insects are occasional problems. Effective control measures include Dycarb (bendiocarb), Dursban (chlorpyriphos), Orthene (acephate), or Enstar (S. kinoprene).

Caterpillars frequently cause a straight line of holes in the leaf when feeding as the leaf unfurls. Large droppings are also observed. Sprays of Orthene (acephate), Dursban (chlorpyriphos), or Dipel *(Bacillus thuringiensis)* help. Snail injury is frequently confused with caterpillar injury, but snails leave squiggly droppings and can usually be found on the soil surface or underneath the pot. Metaldehyde baits can help, and Mesurol may be labeled in some countries.

A somewhat unique pest of calatheas is the microscopic mite *Steneotarsonemus furcatus*. The symptoms are water-soaked lesions or necrotic lines near the leaf margin. This can easily be confused with disease or fluoride injury. Thiodan (endosulfan) is the most effective control to date.

DISORDERS

Fluoride toxicity is a common problem with most *Calathea* varieties. It shows up as a tip or marginal burn, affecting older leaves the most. There may be a yellow halo around the necrotic edge. High light and high temperatures tend to aggravate the problem. You must use low-fluoride phosphate sources and preferably low-fluoride irrigation water. Keeping soil calcium and pH levels up will help, and avoid heat and moisture stress at all costs. Many fluoride-laden calatheas have ended up in the dumpster.

Chlorosis of new foliage is usually caused by low nitrogen or iron. Root rot symptoms may actually be caused by elevated soluble salts instead of disease.

TRICKS

Use low-fluoride liming materials in the media when growing calatheas. Incorporation of gypsum reduces fluoride availability. *C. makoyana* is more tolerant of fluoride than *C. insignis*. *C. rosea-picta* and *C. louisae* are relatively resistant to *Helminthosporium*.

Calatheas like air movement, which tends to increase their leaf size. Irrigate these plants strictly in the middle of the day to help combat disease and fluoride injury. Do not let humidity fall below 60%.

INTERIOR CARE

Calatheas tolerate lower light than many foliage plants. The minimum is about 150 f.c. (1.6 klux). Color improves as light levels of 400 f.c. (4.3 klux) are obtained.

Low temperature of 55F (13C) will cause damage, so avoid drafty areas. Look for plants grown in a high-quality, well-drained potting soil, and you will have much better longevity. A forgiving mix allows you to moisten the medium well; then let the plant dry down somewhat before watering. Do not allow water to remain in saucers. Apply fertilizer very sparingly, preferably from soluble 24-8-16. Overwatering and elevated soluble salts are the most common reasons for plant loss in *Calathea*.

REFERENCES

Bailey, L.H., and E.Z. Bailey. 1976. *Hortus Third*, 198-199. New York: Macmillan.

Chase, A.R. 1986. Calathea. *Foliage Digest* (December): 1-4.

Chase, A.R. 1986. Fertilizer Level Does Not Affect Severity of *Helminthosporium* Leaf Spot of Calatheas. *Foliage Digest* (December): 8.

Chase, A.R. 1987. Susceptibility of Calatheas to *Helminthosporium* Leaf Spot. *Folage Digest* (November): 3.

Chase, A.R. 1989. Fertilizing Calatheas With Slow-Release Fertilizer. *Foliage Digest* (October): 1-3.

Conover, C.A., and R.T. Poole. 1984. *Relationships of Dolomite and Superphosphate to Production of* Calathea. AREC-A Research Report RH-84-16: 1-2. Apopka: Univ. of Florida Agricultural Research and Education Center.

Conover, C.A., and R.T. Poole. 1985. Growth of *Calathea makoyana* as Influenced by Media, Fertilizer and Irrigation. *Nurserymen's Digest* (February): 68-70.

Halawi, M. 1992. *Calathea crocata. GrowerTalks* (August): 19.

Henny, R.J. 1985. Calatheas and Marantas. *Foliage News* 10 (4): 1-4.

CAST-IRON PLANT

(See Other Foliage: Aspidistra*)*

CHAMAEDOREA
Parlor and Bamboo Palms

HABITAT

The genus *Chamaedorea* is aptly named, as it comes from the Greek words for "dwarf" and "gift." These plants are typically small, clumping palms with slender trunks, which makes them attractive, durable interior plants. Due to their popularity as indoor plants, chamaedoreas have been known for many years as parlor palms.

Chamaedoreas come almost exclusively from the rain forests of Central America. They grow in moist, humid jungles, shaded by the larger rain forest plants. Because of this, these palms require low light levels and have relatively weak root systems.

USES

Chamaedorea (pronounced "ka-mie-do-ree-a") is grown with numerous seedlings per pot in containers ranging from 3 to 21 inches (7.5 to 52.5 cm).

Most production of the larger varieties is in 10- to 14-inch (25- to 35-cm) containers. These are usually maintained as houseplants or accent plants in tropical interiorscapes. Smaller varieties, such as *C. elegans,* the Neanthe bella palm, are also produced in smaller containers and even for dish gardens.

Fig. 10 A 10-inch *Chamaedorea elegans* Neanthe Bella. (Photo by the author)

VARIETIES

Chamaedorea elegans, a native of Mexico and Guatemala, has been a versatile interior variety in the trade for many years. The bamboo palm, *C. erumpens,* comes from Belize and Guatemala. This popular indoor palm has somewhat broader pinnae (leaflets) than otherwise similar palms. The name comes from the bamboolike nodes on the trunk. *C. seifrizii* is similar to *C. erumpens* but with more narrow leaves. It also comes from the Yucatan, but it can be grown in higher light. Its common name is the reed palm. Seedlings originating from stock plants in Florida are frequently called Florida Hybrid, as they usually contain genetic elements of both *C. seifrizii* and *C. erumpens.* It has been suggested that *C. seifrizii, C. erumpens,* and Florida Hybrids are all actually one species, *C. seifrizii.* This may be true, but the distinction in the trade will likely persist.

   C. cataractarum has more of an emerald green color and looks rather similar to the areca palm, *Chrysalidocarpus lutescens.* Other varieties occasionally grown in the trade include *Chamaedora geonomiformis, C. metallica, C. tepejilote, C. ernesti-augusti,* and the cold-tolerant *C. microspadix.*

PROPAGATION

Almost all chamaedoreas are propagated from seed, which is typically collected in late summer either in Central America or Florida. Fresh seed is best, but seed can be stored for a month or so unrefrigerated in sealed bags. Seed should be sown shallowly and at consistent depth for even germination. Most planting is done from September to November. Some growers like to seed directly, using around 20 seeds for a 4- to 6-inch (10- to 15-cm) pot.

Seed can take up to a year to germinate, but it will come up in two to four months if grown with bottom heat. The best chamaedorea propagators I have seen grow in flats with bottom heat from electric cable or hot water pipes. The target soil temperature should be 90F (32C). Drop this back to 80F (26.5C) after most of the seeds have germinated, or root growth will be stunted.

When using flats many growers transplant the seedlings right after germination, before the initial shoot has opened into a leaf. At least eight to 12 seedlings are evenly spaced in either a 6-inch (15-cm) or a 10-inch (25-cm) container. Another method is to pot five plants in a 6-inch pot, then later pot three of those in a 10- or 14-inch (35-cm) container, resulting in 15 well-spaced plants per pot. Pot them full to reduce the cull count later.

CULTURE

Many growers like to start their chamaedoreas (except *C. elegans*) in full sun to get better stem caliper and basal shoots. A seedling should be starting on its third leaf before going into the sun, or else strong sun can kill the seedling. Starting in full sun gives you pale color and leaf necrosis, but you ultimately get a stronger plant. Greenhouse growers may prefer to start chamaedoreas in about 30% shade after potting. More mature plants are usually grown at 73% shade, sometimes 80%, especially for *C. elegans*. These light levels are approximately equivalent to 1,500 to 3,000 f.c. (16.2 to 32.4 klux). *Chamaedoreas* don't do well in highly absorbent potting media. Mixes emphasizing sphagnum peat, perlite, and sand are preferred. Pine bark and other wood products are acceptable if used in moderation. A well-drained mix with reduced moisture-holding capacity is best, and these palms should be watered conservatively. A major mistake of many *Chamaedorea* growers is overwatering.

Preferred temperatures are 75 to 90F (24 to 32C). Root activity decreases when soil temperature reaches 65F (18C). These palms tend to suffer injury if temperatures drop below 45F (7C). Plants should be finished for at least 10 weeks in 73 to 80% shade for good interior performance. Ronstar (oxadiazon) works well as a preemergent herbicide.

NUTRITION

The seedlings are rather delicate, so try to be gentle when fertilizing young chamaedoreas. A 3-1-2 ratio of N-P$_2$O$_5$-K$_2$O is popular. Small pots are frequently grown with liquid feed of 24-8-16 or 20-10-20 at 200 ppm nitrogen. Coated, slow-release fertilizers are also very useful for these slow-growing palms. Research has shown that the source of nitrogen is not critical, at least under tropical conditions. Chamaedoreas have weak root systems, so it is helpful to drench the pots at planting with fungicides and a high-phosphate starter fertilizer in order to stimulate root growth.

Trace element deficiencies are rather rare in these palms. They occasionally show iron deficiency as chlorosis in the new leaves, especially when soil pH is high or soil remains waterlogged. Lack of zinc results in smaller leaves than normal. Many growers occasionally spray a trace element preparation to help avoid deficiency problems.

TABLE 6 Leaf analysis rating standards for *Chamaedorea erumpens*

Nutrient (%)	Very low	Low	Medium	High	Very high
Nitrogen	<2.0	2.0–2.4	2.5–3.5	3.6–4.5	>4.5
Sulfur	<0.15	0.15–0.20	0.21–0.75	0.76–1.00	>1.00
Phosphorus	<0.11	0.11–0.14	0.15–0.30	0.31–0.75	>0.75
Potassium	<1.20	1.25–1.55	1.60–2.75	2.80–4.00	>4.05
Magnesium	<0.21	0.21–0.24	0.25–0.75	0.76–1.00	>1.00
Calcium	<0.40	0.40–0.99	1.00–2.50	2.51–3.25	>3.25
Sodium			<0.11	0.11–0.25	>0.25
(ppm)					
Iron	<40	40–49	50–300	301–500	>500
Aluminum			<101	101–500	>500
Manganese	<40	40–49	50–250	251–500	>500
Boron	<18	18–24	25–60	61–100	>100
Copper	<4	4–5	6–50	51–200	>200
Zinc	<18	18–24	25–200	201–400	>400

Sources: Institute of Food and Agricultural Sciences, Apopka, Florida; Dr. Benjamin Wolf, Fort Lauderdale, Florida.

Notes: Common names include bamboo palm, parlor palm, *seifrizii* palm. Sample of most recent fully mature leaves, no petioles.

DISEASES

Chamaedoreas are rather disease prone, especially when grown under shade-house or outdoor conditions. The major disease problem is *Gliocladium*, commonly known as pink rot. Plants display a yellowing of the older fronds, with numerous fruiting bodies (conidia) on the old leaf sheaths. These conidia may be pink, orange, or white. A black substance may ooze from lesions in the basal stem. The fungus grows better in cool weather and is somewhat less of a problem above 86F (30C). *Gliocladium* is a good saprophyte, so it likes to grow on decaying older leaves and sheaths. Periodically trimming and removing old, necrotic leaves is helpful, but be careful not to injure stem tissue, or the whole stem could die. Sprays of Daconil (chlorothalonil) or Cleary's 3336 (thiophanate methyl) plus Dithane (mancozeb) are helpful.

Phytophthora stem rot is frequently found in plants also displaying *Gliocladium* symptoms. The roots turn black, and the stem will turn dark from the bottom, with the necrosis working its way upward. This frequently happens when plants are overwatered or exposed to frequent rainfall. Keeping the medium dry helps, as do soil drenches of Subdue (metalaxyl) or applications of Aliette (fosetyl-aluminum).

 Phosphorus leaches from most potting soils.

Damping-off, a common problem in seedlings, is caused by several different fungi. In early stages it is useful to drench either with Banrot (etridiazole plus thiophanate methyl) or with Subdue (metalaxyl) combined with either a thiophanate methyl fungicide or Chipco 26019 (iprodione). Soluble, high-phosphate starter fertilizers also help to generate or regenerate roots.

Fusarium can cause a rather severe seed rot. If seeds show fungal decay, a drench with thiophanate methyl or iprodione can be beneficial. *Cylindrocladium* can cause a fairly severe leaf spot. Numerous brown spots, either round or oval in shape, are seen, usually with chlorotic halos around them. Sprays of Chipco 26019 (iprodione) are helpful. Many growers apply occasional preventive sprays with either Chipco 26019, Daconil (chlorothalonil), or Dithane (mancozeb).

INSECT AND MITE PESTS

Spider mites (technically not insects) are the major pest of virtually all chamaedoreas grown in nurseries. The two-spotted mite population can build up quickly under warm greenhouse conditions. Growers usually have to spray preventatively with Avid (abamectin), Pentac (dienoclor), or Ornamite (propargite) to avoid severe outbreaks. Fungus gnats can be a problem in moist seed beds. Applications of Gnatrol (*Bacillus thuringiensis* var. *Israelensis*) are useful, as is Resmethrin or Azatin (azadirachtin).

Root mealybugs (*Rhizoecus* or *Geococcus*) can feed on roots and cause lack of plant vigor. The nymphs spread from pot to pot, floating on water and entering through drain holes. Diazinon drenches are helpful, though they can occasionally be phytotoxic. When snails are a problem, metaldehyde baits are somewhat helpful, as is mechanical control.

Ambrosia beetles (*Xyleborus*) can cause very serious losses in seed beds. The insect, which looks like a tiny cockroach, bores into the seed, leaving a hole about the size one would make with a thumbtack. Watch carefully for this problem, and if you encounter it, Lindane may be helpful if registered for your situation. A tarsonemid mite occasionally causes leaf distortion in parlor palms. Applications of Thiodan (endosulfan), with good coverage and a wetting agent, help control it.

A fairly recent serious problem of chamaedoreas involves the banana moth (*Opogona*). When experiencing chamaedorea stem dieback, take a pocketknife and slit the decaying stem from the base toward the top. Inside you may find a transparent, active, highly segmented caterpillar. It will have a dark head and be anywhere from one-half to $1^1/_2$ inches (1.25 to 4 cm) long. The banana moth can cause serious losses, though sprays of Sevin (carbaryl), Lindane, or the beneficial nematode Steinernema (Biovector) are effective.

DISORDERS

Tip burn is a common disorder of chamaedoreas when the soil is poorly aerated and waterlogged, when significant root loss has occurred, or when soluble salts are above 1,000 ppm. Improving the roots with fungicide drenches and getting the salts down will reduce the problem. Fluoride and boron toxicity also cause tip burn in these palms. The two symptoms are similar, though excess boron tends to have a yellow halo between the brown necrotic tissue and the green tissue. Keeping the pH and calcium levels up helps reduce toxicity symptoms from both of these elements. As a chamaedorea gets older in the pot, the physical structure of the medium breaks down, and this contributes to the various forms of tip burn, as well.

Chamaedoreas also tend to drop the leaves nearest the inflorescence. This is presumably to aid in pollination. If your chams are flowering, there is very little you can do about the leaf drop, though some growers cut the bloom spikes off. Also, when growing outdoors in shadehouses, count on losing two to three plants per pot to disease or insects.

TRICKS

Chamaedoreas are among the very few palms that can actually be cut back, if necessary. The bamboo palm varieties can be cut back to within 2 inches (5 cm) of the soil line, and they will send back multiple basal shoots. Sprays of Aliette at 5 pounds per 100 gallons (2.3 kg per 400 l) induce additional basal shoots, especially in bamboo palm varieties.

Light is not required for seed germination in these palms. Wear gloves when handling *C. seifrizii* seeds because the calcium oxalate crystals they generate can be irritating. Don't soak the seeds in giberellic acid because the seedlings will stretch. Seed beds under bottom heat dry out frequently, so don't forget to monitor the soil moisture.

Leave pine bark out of the potting mix if you plan to ship to Europe. Space the seedlings evenly in the pot, and try to use a potting mix made at least 60% of ingredients that will not physically degrade over time.

INTERIOR CARE

Chamaedoreas are quite durable under low light conditions, but they need at least 75 to 150 f.c. (0.8 to 1.6 klux) of light to retain quality. A north window exposure without direct sunlight is popular. Interior bamboo and parlor palms prefer 40 to 60% relative humidity. Try to keep the soil evenly moist, but absolutely do not overwater. Fertilize very lightly, perhaps every four to five months, with 1 teaspoon of 20-20-20 per gallon. Because of the weak root systems, repotting is generally not needed, but if the media mix is broken down and muddy or mucky in appearance, placing chams in fresh media is desirable.

REFERENCES

Burch, D., R. Atilano, and J. Reinert. 1984. *Indoor Palm Production Guide for Commercial Growers.* IFAS Ornamental Horticulture Fact Sheet OHC-8: 1-6. Gainesville: Universityof Florida Cooperative Extension Service, Institute of Food and Agricultural Sciences.

Chase, AR. 1993. Common Diseases of Palms. *Southern Nursery Digest* (March): 20-21.

Conover, C.A., and R.T. Poole. 1986. Effects of Nitrogen Source and Potting Media on Growth of *Chamaedorea elegans, Dieffenbachia maculata* 'Camille' and *Peperomia obtusifolia. Proc. Fla. State Hort. Soc.* 99: 282-284.

Donselman, H., and T.K. Broschat. 1986. Phytotoxicity of Several Pre- and Postemergent Herbicides on Container Grown Palms. *Proc. Fla. State Hort. Soc.* 99: 273-294.

Griffith, L.P. 1983. Growing Chamaedoreas. *Florida Nurseryman* (September): 73-74.

Leahy, R.M. 1990. Cylindrocladium Leaf Spot of Palms. *Foliage Digest* (April): 3.

Osborne, L.S., A.R. Chase, and O.G. Burch. 1983. *Chamaedorea Palms.* ARC-A Foliage Plant Research Note RH-1983-E. Apopka: University of Florida Agricultural Research and Education Center.

Poole, R.T., and C.A. Conover. 1974. Germination of Neanthe Bella Palm Seeds. *Proc. Fla. State Hort. Soc:* 87: 429-430.

CHENILLE PLANT

(See Other Foliage: Acalyphas*)*

CHINA DOLL

(See Other Foliage: Radermachera*)*

CHINESE EVERGREEN

(See Other Foliage: Aglaonema*)*

CHLOROPHYTUM
Spider Plant

HABITAT

Chlorophytums are commonly known as spider plants, sometimes as airplane plants or ribbon plants. Spider plants are distinguished from most other ornamental plants by their numerous stolons, which form clusters of hanging plantlets that resemble spiders.

The 215 species come from all parts of the world except for North America and Europe. The principal chlorophytum in the trade, *C. comosum,* is native to South Africa. It grows on forest floors in the eastern part of the country, which has very distinct wet and dry seasons.

Fig. 11 Variegated *Chlorophytum comosum* in Delray Beach, Florida. (Photo by the author)

USES

Because of the plant's arched leaves and trailing habit, virtually all chlorophytum production is in hanging baskets, the majority in 6-, 8-, and 10-inch (15-, 20-, and 25-cm) basket containers. I have occasionally seen chlorophytums used as ground-cover plantings in tropical landscapes and in interiors. Spider plants serve well when associated with rock formations or waterfalls, though they can be sensitive to chemicals in waterfalls and fountains. Chlorophytums are known to be one of the best plants for helping to purify the air in interior environments. They require minimal care and make durable houseplants.

VARIETIES

Chlorophytum comosum cv. Vittatum is by far the most commonly produced variety, making up about 90% of spider plant sales in most markets. It has long, tapered, medium green leaves with a large, central, white stripe. Vittatum is also one of the larger spider plant cultivars. *C. comosum* cv. Variegatum is a smaller version of variegated spider plant, with white bands along the edges of the leaves and green in the middle. The less common *C. comosum* cv. Mandaianum looks something like the reverse of Variegatum, in that this is also a smaller variety but with green leaf margins and a bright yellow stripe along the midrib.

The solid green spider plant is *C. comosum*, not *C. capense*, as had been recorded in several early research papers. It has medium to lime green foliage without variegation and produces numerous stolons and plantlets.

PROPAGATION

Spider plants can be grown from seeds, but almost all production is from plantlets. Most nurseries do not maintain stock plants, instead taking the plantlets from the ends of runners from existing plants in production. Plantlet production is distinctly photoperiod related. The variegated varieties tend to produce more runners during short days (winter), while the solid green variety creates more runners during long days (summer).

Growers typically put four plantlets in a 6-inch (15-cm) basket, five in an 8-inch (20-cm) basket, and six in a 10-inch (25-cm) basket. Plantlets are watered in at potting but not misted. Baskets can be hung on pipes or hooks in the greenhouse at this time, though many growers prefer to keep the plants pot to pot on a bench for the first four to six weeks to save space.

CULTURE

Chlorophytums are normally grown with between 1,000 and 2,500 f.c. (10.8 to 27 klux). That requirement will vary somewhat with cultivar and time of year. Higher light levels cause bleaching of foliage and tip burn. When light levels are too low, the foliage droops severely, and very little growth occurs. Flowering and plantlet production are greater in brighter light.

Potting media vary somewhat, as spider plants are a fairly short-term crop and tolerant of many types of soils. A typical mix would include 50% peat, with the balance made up of various combinations of pine bark, vermiculite, perlite, or styrofoam beads. In order to remove fluoride from perlite, it is helpful to leach it thoroughly once prior to blending.

Preferred temperatures are between 65 and 90F (18 and 32C). Growth tends to slow below 65F (18C). Growers in northern climates may keep their night temperatures as low as 55F (13C), which will save on the heating bill but will slow growth somewhat, as well. Spider plants usually tolerate down to 35F (2C) before cold damage is observed. Chlorophytums are one of the easier foliage plants to grow, but you should not try them if you have poor quality irrigation water.

NUTRITION

The nutrient requirements of spider plants are lower than many foliage plants. They actually get by on relatively little fertilizer, and I know growers who virtually do not fertilize them! Chlorophytums have very long, tapered leaves; therefore, any excess elements taken up in the transpiration stream will tend to migrate ultimately to the leaf tip, causing injury. Fluoride, boron, and sodium can all cause tip burn in spider plants, either individually or in combination. I do not recommend boron in the fertilizer program for spider plants and try to use low-fluoride irrigation water and fertilizers.

Many growers use light rates of soluble 20-10-20, at perhaps 150 ppm nitrogen two to three times per week. Other growers prefer a coated, slow-release fertilizer, generally lightly top-dressed at the low or medium label rate. It is usually not necessary to incorporate minor elements into the media for chlorophytums, but if you do, try to use a trace element preparation without boron.

DISEASES

For once, here's a foliage plant whose disease problems are relatively rare. Most foliage plants, as you have read, require significant disease control efforts, but with this one, as long as you don't overwater, disease problems are unlikely. Reported pathogens include leaf spots caused by *Alternaria*, *Cercospora*, *Fusarium*, and *Phyllosticta*. Reported root rots include *Pythium*, *Rhizoctonia*, and *Sclerotium rolfsii*, or southern blight. *Erwinia* and *Pseudomonas* are reported to cause occasional bacterial leaf spots and blights. Pathogen problems in spider plants are so rare that I will not bother describing them or

TABLE 7 Leaf analysis rating standards for *Chlorophytum comosum*

Nutrient (%)	Very low	Low	Medium	High	Very high
Nitrogen	<1.1	1.1–1.4	1.5–2.5	2.5–3.5	>3.5
Sulfur	<0.14	0.14–0.19	0.20–0.60	0.61–0.75	>0.75
Phosphorus	<0.07	0.07–0.09	0.10–0.30	0.31–0.50	>0.50
Potassium	<2.80	2.8–3.4	3.5–5.0	5.1–6.0	>6.0
Magnesium	<0.25	0.25–0.29	0.3–1.0	1.1–1.5	>1.5
Calcium	<0.70	0.70–0.90	1.0–2.0	2.1–3.0	>3.0
Sodium			<0.20	0.21–0.50	>0.50
(ppm)					
Iron	<35	35–49	50–200	201–500	>500
Aluminum			<250	251–1000	>1000
Manganese	<35	35–49	50–200	201–500	>500
Boron	<15	15–19	20–50	51–75	>75
Copper	<5	5–7	8–100	101–500	>500
Zinc	<16	16–19	20–200	201–500	>500

Sources: Institute of Food and Agricultural Sciences, Apopka, Florida; Dr. Benjamin Wolf, Fort Lauderdale, Florida.
Notes: Common names include spider plant, airplane plant. Sample of most recent fully formed leaves, no petoles.

their control measures. If something pops up, you can get the control information from other disease sections in this book. Most problems in spider plants are not disease related.

INSECT AND MITE PESTS

Caterpillars can be a significant problem in chlorophytum production, especially during spring and summer months. They create large chewed areas on the foliage, frequently starting at the leaf tip and working back toward the center of the plant. When you encounter it, sprays of Orthene (acephate) or one of the *Bacillus thuringiensis* insecticides should control them.

   Both foliar and root mealybugs can occasionally become a problem, usually on one or a few isolated plants. Discard the heavily infested baskets and spray the rest with either Dycarb (bendiocarb), Enstar (S-kinoprene), or

Orthene (acephate). Root mealybugs are usually treated with a Diazinon drench. Scale insects are also an occasional problem on spider plants. Several types of scale insects are reported. The control measures are generally the same as for mealybugs.

Snails are sometimes a problem, leaving chewing injury similar to that of caterpillars, but with thin, squiggly droppings rather than the larger, round droppings left by caterpillars. Keeping the baskets hung up and off the benches will help reduce snail trouble.

DISORDERS

By far the major disorder encountered by chlorophytum growers is tip burn. This is usually caused by accumulation of fluoride, boron, or sodium. Be aware, however, that temperatures above 90F (32C) also contribute to tip burn, especially when fluoride, boron, or sodium are elevated. Fluoride injury symptoms are generally more reddish in appearance, whereas burned tip tissue is more tan to gray when caused by boron accumulation. Elevated sodium tends to give you black leaf tips.

Spider plants in general are highly reactive to fumes in the atmosphere. The insecticide Dursban (chlorpyriphos) is reported to be phytotoxic on spider plants, so don't use it to control scale, mealybugs, or southern blight. Chlorophytums develop epinasty (severe drooping) when exposed to 5 ppm ethylene or greater for any period of time. The drooping usually goes away once the ethylene dissipates. However, don't confuse this with droop caused by low light. Chlorophytums are also sensitive to bleach fumes, so be careful about having them in close proximity when bleaching benches and sidewalks.

TRICKS

First of all, try to use the highest-quality water you can, as low as possible in boron, fluoride, and sodium. Do not use superphosphate as a phosphorus source in the fertilizer program. Chlorophytums appear to like water on their leaves from time to time, perhaps for its cooling effect. If you are growing with drip irrigation, it helps to get the leaves wet with overhead irrigation or occasional hand watering, especially in warm periods. In hot weather irrigate in the middle part of the day to increase humidity and cooling. Use lime sources low in fluoride in your media. Incorporating gypsum (calcium sulfate) into the potting media at 3 pounds per cubic yard (1.8 kg per m³) will help bind fluoride and reduce sodium uptake. If tip burn is a chronic problem for you, some of it may be alleviated with antitranspirant sprays about every three weeks.

Spider plants bloom better under warmer temperatures. Runner and plantlet production are significantly increased if you utilize night interruption. Turn on lights such as mum lights for four hours during the middle of the night.

INTERIOR CARE

When reasonably cared for, spider plants last very well in interior environments. They survive at a minimum of 75 to 100 f.c. (0.8 to 1.1 klux), though for best quality they should have at least 150 to 200 f.c. (1.6 to 2.7 klux). Irrigation requirements are moderate, but don't let a plant dry out completely, or tip burn could be worsened. Do not use fluoridated tap water on spider plants. Rainwater or well water is preferred because of the fluoride problem.

Feed spider plants very sparingly, only about one-half teaspoon of soluble 20-20-20 or 20-10-20 per gallon, and preferably not more often than every three or four months. They will get pot bound in time in smaller baskets; repotting with fresh media will result in larger plants. New plants can be created rather easily from the hanging plantlets.

Fluoridated city water usually has 1 ppm fluoride, four times the amount considered safe for sensitive plants.

REFERENCES

Bailey, L.H., and E.Z. Bailey. 1976. *Hortus Third*, 265. New York: Macmillan.

Blessington, T.M., and P.C. Collins. 1993. *Foliage Plants*, 59-61. Batavia, Ill.: Ball Publishing.

Conover, C.A., A.R. Chase, and L.S. Osborne. 1987. *Spider Plant. Foliage Digest* 10 (7): 1-3.

Hammer, P.A., and G. Holton. 1974. Asexual Reproduction of Spider Plant, *Chlorophytum elatum*, by Day Length Control. *Florist's Review* (September 4): 35, 76.

Heins, R.D., and H.F. Wilkins. 1978. Influence of Photoperiod and Light Quality on Stolon Formation and Flowering of *Chlorophytum comosum* (Thunb.) Jacques. *J. Amer. Soc. Hort. Sci.* 103 (5): 687-689.

Obermeyer, A.A. 1958. A Revision of the South African Species of *Anthericum, Chlorophytum* and *Trachyandra. Bothalia*, pt.1, 669-700. Pretoria, South Africa: The Government Printer.

Sheely, H.C., and J.W. White. 1980. Factors Affecting Lamina Necrosis of *Chlorophytum comosum* Thunb. *Hortscience* 15 (4): 502-504.

Watkins, J.V., and T.J. Sheehan. 1975. *Florida Landscape Plants,* 97. Gainesville: The University Presses of Florida.

CHRYSALIDOCARPUS

Areca Palm

HABITAT

The areca palm is among the most popular of the ornamental palms in the world. Literally millions of areca palms are produced in containers every year, yet at one time this palm was placed on the endangered species list. the *Chrysalidocarpus lutescens* was formerly known as the *Areca lutescens*, hence the common name. The

Fig. 12 The popular areca palm, seen here in 14-inch pots. (Photo by the author)

palm is also known by several other common names, but the name "areca palm" seems to be familiar throughout most of the world. *Chrysalidocarpus* is Greek for "golden fruit," and *lutescens* means "becoming yellow," both of which are appropriate names because areca palms tend to become chlorotic.

C. *lutescens* is a pinnate, clustering palm native to Madagascar. It is said to grow to 30 feet (9 m), but in most cases mature specimens in the landscape are 12 to 15 feet (3.5 to 4.5 m). Arecas do better at lower altitudes with warmer night temperatures and are typically found in USDA Zone 10 in the tropics, occasionally in protected areas of USDA Zone 9.

USES

The golden stems and petioles of areca palms, along with their gracefully arching fronds, make them attractive as interior specimens as well as landscape plants in tropical areas. Despite their popularity, areca palms are not particularly good houseplants. The most common type of areca produced is a 10-inch (25 cm) container with 10 to 30 plants per pot. These are largely shade grown and sold at about 36 inches (0.9 m) high. Also common are 6- and 8-inch (15- and 20-cm) containers, as are large specimens in containers of 24 inches (60 cm) and larger. These palms are virtually always grown with many plants per pot, giving them a full, somewhat grassy appearance. Ten-inch (25-cm) areca palms are popular office and houseplants in brighter interior locations. Larger specimens are popular for such open indoor spaces as shopping malls, hotel atriums, and airports.

VARIETIES

C. *lutescens* is the only widely cultivated plant in the genus. Palms grown from Madagascar seed are visually identical, though in my experience they are more cold sensitive than arecas grown from Florida or Central American seed. About 20 species of *Chrysalidocarpus* exist, but the only other one occasionally seen in commercial nurseries is the cabada palm, C. *cabadae*. The betel nut palm, *Areca catechu*, is a relative of C. *lutescens* but no longer in the same genus. Occasionally grown as a single-stem foliage plant, the betel nut palm is completely different in appearance from the common areca palm.

Many monocots, such as palms and dracaenas, can absorb iron in unavailable forms.

PROPAGATION

Areca palms are propagated from seeds, which mature between July and September in Florida, earlier in Central America and elsewhere. The seed is about the size of a jelly bean and numbers about 50 to the ounce (0.57 g each). As the seed matures, the pulp changes from green to yellow-green to orange. Usually the more orange in color, the higher the germination percentage. Green seeds tend not to germinate well if the pulp is removed prior to planting. Leaving the pulp on green seed improves germination, but the seed must be planted quickly. Areca seed can be stored up to a year if it is cleaned of pulp, air dried, and treated with a fungicide, such as Thiram. Store the seed in an airtight container, such as a thick garbage bag, at 73F (23C).

It is important to plant areca seeds all at the same depth, or germination will be irregular. Irregular germination results in some seedlings in the pot being thick, with heavy stem caliper, while the seedlings coming up later cannot compete and will stay very thin. European markets seem to prefer more caliper in areca stems, whereas the United States market prefers more uniform stem caliper, more of a grassy appearance.

Optimum soil temperature for germination is 85 to 95F (30 to 35C). Bottom heat is not generally required in warm climates. Some people plant seed in open beds, while others plant directly in 6- or 10-inch (15- or 25-cm) containers. The best method, in my experience, however, is to plant seed in liner trays, preferably ones with 3- to 4-inch (7.5- to 10-cm) cells. The seeds should be barely but uniformly covered. Germination takes around six weeks, but soaking seed in water for a day or two before planting hastens germination.

CULTURE

Most arecas are produced in peat-based mixes, either a reed sedge or sphagnum peat or a combination, usually mixed with bark, perhaps wood shavings, and about 10% sand. Because areca palms are a relatively long crop (about 18 months for a 10-inch, or 25-cm, pot), it is important to have a potting mix that will retain good physical structure for the long term. A pH of 6 to 6.5 is preferred. Areca palms may respond to liming when the pH drops below 5.7. It is helpful to incorporate a trace element preparation into the potting medium, and many growers spray such preparations on a regular basis to maintain color. Overwatering weakens roots and reduces color, so let the soil dry between waterings.

Books typically say to grow areca palms at 5,000 to 6,000 f.c. (54 to 64.8 klux), or around 55% shade. Most growers do better, however, growing in 63 or 73% shade, closer to 3,500 f.c. (37.7 klux). Arecas have a distinct tendency to turn yellow; growing them under lower light helps to maintain color. Some growers start them in full sun, then move them into shaded conditions when about half grown. Arecas grow faster if they get some degree of shade when they are young, at least in hot climates.

Ideal soil and air temperatures for growing areca palms are 70 to 80F (21 to 27C). It is best not to have air temperatures exceed 95F (35C), though in tropical shadehouses temperatures sometimes go higher. The palms tolerate temperatures down to about freezing, but as temperatures approach freezing for any length of time, the plants may be injured. The chloroplasts can be injured at low temperature, causing the plants to turn somewhat orange or even necrotic. It is more difficult to keep arecas green in winter.

Nutrition

Most areca growers use a 3-1-2 ratio of N-P_2O_5-K_2O. Some top-dress periodically with granular fertilizers, while others top-dress or incorporate such slow-release fertilizers as 19-6-12 or 18-6-8. Arecas are moderate to heavy feeders, so use the medium to high fertilizer rates but no more. Constant liquid feed from soluble 24-8-16 or 20-20-20 is also used, especially for smaller pot sizes and liners.

Areca palms have rather high requirements for magnesium and trace elements. The books (except this one) advise not to spray iron on areca palms, or leaf spotting could develop. The truth is that most growers need to spray trace element preparations containing iron and manganese on a regular basis to help maintain good leaf color. It is common to add urea or potassium nitrate to the trace element spray, as it is thought to aid in absorption. Foliar sprays of magnesium from sulfate, nitrate, or chelated sources tend to improve color, as well, especially at higher light levels. Once areca palms start to turn yellow, it can be difficult to bring the color back. Therefore, many growers like to use periodic nutritional sprays as supplements.

TABLE 8 Leaf analysis rating standards for *Chrysalidocarpus lutescens*

Nutrient (%)	Very low	Low	Medium	High	Very high
Nitrogen	<2.0	2.0–2.4	2.5–3.5	3.6–4.5	>4.5
Sulfur	<0.15	0.15–0.20	0.21–0.75	0.76–1.25	>1.25
Phosphorus	<0.11	0.11–0.14	0.15–0.75	0.76–1.25	>1.25
Potassium	<1.01	1.01–1.25	1.30–2.75	2.80–4.00	>4.05
Magnesium	<0.21	0.21–0.24	0.25–0.75	0.76–1.00	>1.00
Calcium	<0.40	0.40–0.99	1.00–2.50	2.51–3.25	>3.25
Sodium			<0.11	0.11–0.25	>0.25
(ppm)					
Iron	<40	40–49	50–300	301–1000	>1000
Aluminum			<251	251–2000	>2000
Manganese	<40	400–49	50–250	251–1000	>1000
Boron	<12	12–14	15–60	61–100	>100
Copper	<4	4–5	6–50	51–500	>500
Zinc	<18	18–24	25–200	201–1000	>1000

Sources: Institute of Food and Agricultural Sciences, Apopka, Florida; Dr. Benjamin Wolf, Fort Lauderdale, Florida.

Notes: Common name: areca palm. Sample of most recent fully mature leaves, no petioles.

DISEASES

By far the major disease problem with areca palms is the *Helminthosporium* complex. Several different fungi may be involved, including *Bipolaris, Drechslera, Exerohilum,* and *Phaeotrichoconis.* Plant pathologists can tell these apart, but to the grower they all form annoying small, circular, brown spots. The spots can be very numerous, and under intense disease pressure, entire emerging fronds can be killed. Controls include good nutrition, keeping the foliage dry, and regular preventive sprays with Daconil (chlorothalonil), Dithane (mancozeb), Chipco 26019 (iprodione), Ornalin (vinclozolin), or Carbamate (ferbam).

Gloeosporium, another common fungal leaf spot, is similar in appearance to the *Helminthosporium* complex, but the spots may be larger and more oval. Controls are similar, though Cleary's 3336 or Domain (thiophanate methyl) are also effective against this fungus. Banner (propiconazole) sprays have been used to control both *Helminthosporium* and *Gloeosporium*.

Several fungal root diseases are common, especially in wet weather or with poorly drained media. *Pythium, Fusarium, Rhizoctonia,* and *Phytophthora* all may cause root rot. Soil drenches with thiophanate methyl and Subdue (metalaxyl) along with careful watering are helpful. There are no reported bacterial or viral diseases of areca palm.

INSECT AND MITE PESTS

Spider mites, usually the two-spotted mite *(Tetranychus)*, are the number-one pest problem. Mite feeding causes small, white specks on foliage and leaf discoloration. Severe infestations cause death of entire leaves. The juvenile foliage of young arecas is most susceptible. The mature fronds that begin to develop once arecas reach 4 feet (1.2 m) are thicker and more mite resistant. Growers usually spray preventatively with Avid (abamectin), Mavrik (tau-Fluvalinate), Pentac (dienochlor), or insecticidal soaps. Failure to control mites is usually related to inadequate spray coverage.

Caterpillars are an occasional problem. When they do occur, growers usually use Dipel *(Bacillus thuringiensis)*, Sevin (carbaryl), or Orthene (acephate). Foliar and root mealybugs are sometimes encountered. Sprays with Dycarb (bendiocarb) help against foliar mealybugs, while Diazinon drenches are effective against root mealybugs. Scale insects, especially the magnolia or oleander scale, can be a problem. Sprays of Cygon (dimethoate) during cool temperatures will help. Supracide (methidathion) is also very good.

Millipedes can be very destructive, not because they feed on the plants, but because their activities physically degrade the potting medium. The presence of millipedes also generates a distinct, somewhat musty odor. Soil drenches with Diazinon or Sevin (carbaryl) should control them. White grubs occasionally eat all of the roots in a young plant, causing plant death. It is best simply to discard the plant and soil when this happens.

DISORDERS

Imbalances between iron and manganese, resulting in yellow newer leaves, are the most common disorders encountered by areca growers. Distinguishing whether the plants need more manganese or iron is difficult, being best determined through leaf analysis. Be aware that areca palms, like many other monocots, can absorb iron in unavailable forms. Areca palms lacking zinc have small leaves with necrotic tips. Sprays or drenches with the appropriate minor element usually help, though the remedy may take some time.

Fluoride toxicity occasionally causes spots near the midrib and toward the leaf tip. Fluoride toxicity, usually a relatively minor problem in areca palms, can be controlled fairly well with good soil calcium levels. Areca palms are sensitive to soluble salts. Symptoms include a downward curl of the foliage at the leaf tips, as well as leaf tip dieback. Arecas don't take up nitrogen well when salts are high, which is when very few fine hair roots are observed. Leaching with extra irrigation water once or twice usually corrects the problem.

If the potting media physically degrade as plants age, arecas will have increased problems with root rot diseases and nutritional deficiencies, especially those of nitrogen, potassium, and iron. Let the soil dry between waterings as best you can, and try to apply the correct supplemental nutritional sprays. Repotting into new media or larger containers may be necessary. Again, millipedes degrade the media, so you need to control them at all costs to grow arecas successfully.

TRICKS

While the technique is rarely used by growers, soaking areca palm seed in hot sulfuric acid aids in germination. Soaking in giberellic acid prior to planting is discouraged, as seedlings tend to stretch. Seed viability can be estimated quickly and accurately using Tetrazolium. Soaking in 10% Tetrazolium will turn transpiring embryos red, while nonviable seeds don't change color. Some growers like to spray Agrimycin 17 on arecas, but this is a waste of time and contrary to the chemical label. Trace element sprays, especially sprays containing copper, are capable of causing leaf spots. Don't spray copper fungicides or Captan on areca palms.

Well-fed areca palms are more resistant to leaf spot disease than plants with deficiencies. *Helminthosporium* generally sporulates better under high light, and plants in full sun are especially at risk of leaf spot. When plants turn orange from cold injury, repeated sprays with magnesium help turn the orange color back into green. Finally, do not plant areca palms even an inch (2.5 cm) too deeply, or problems with color or growth will occur.

INTERIOR CARE

Arecas are sensitive to soluble salts, so remove any visible granular fertilizer on the soil surface. Try to maintain average temperatures between 60 and 80F (16 and 27C). These palms need relatively bright light, at least 150 to 400 f.c. (1.6 to 4.3 klux). As indicated earlier, areca palms are not particularly long-lasting houseplants, but they will serve well for a time with good light and avoidance of overwatering. Also avoid irrigating with salty water or fluoridated tap water.

Watch for spider mites; spray them with an insecticidal soap or a miticide registered for interior use. Let the potting medium become fairly dry before irrigating. If the mix has broken down and appears muddy, replanting in a slightly larger pot with fresh medium will be helpful.

REFERENCES

Blessington, T.M., and P.C. Collins. 1993. *Foliage Plants: Prolonging Quality*, 62-64. Batavia, Ill.: Ball Publishing.

Burch, D., R. Atilano, and J. Reinert. 1984. *Indoor Palm Production Guide for Commercial Growers*. IFAS Ornamental Horticulture Commercial Fact Sheet OHC-8. Gainesville: University of Florida Cooperative Extension Service, Institute of Food and Agricultural Sciences.

Chase, A.R. 1982. Dematiaceous Leaf Spots of *Chrysalidocarpus lutescens* and Other Palms in Florida. *Plant Disease* (August): 697-699.

Chase, A.R., and R.T. Poole. 1982. *Suggestions for Chemical Control of Areca Palm Leaf Spot*. Agricultural Research Center–Apopka Research Report RH 1982-24. Apopka: Agricultural Research Center.

Chase, A.R., R.T. Poole, and L.S. Osborne. 1985. Areca Palm. *Nurserymen's Digest* (November): 72-74.

Griffith, L.P. 1983. Areca Palm Production. *Florida Nurseryman* (January): 120-121.

Poole, R.T., and C.A. Conover. 1982. *Phytotoxicity of Palms Induced by Foliar Applications of Copper*. ARC-A Research Report RH-82-3. Apopka: Agricultural Research Center.

Schmidt, L., and F.D. Rauch. 1983. *Effects of Presoaking Seed of* Chrysalidocarpus lutescens *in Water and Gibberellic Acid*. Hawaii Institute of Tropical Agriculture and Human Resources Journal Series No. 2713: 4-5. Honolulu: University of Hawaii Institute of Tropical Agriculture and Human Resources.

CISSUS
Grape Ivy

HABITAT

Though it is a member of the grape family, Vitaceae, grape ivy is a true ivy, with the genus name *Cissus* coming from the Greek word for "ivy." There are about 350 species, mostly tropical and subtropical vines. Most grape ivies grow as vines with three leaflets, a compound leaf typically being 2 to 4 inches (5 to 10 cm) long. The curious part

Fig. 13 *Cissus* Ellen Danica, a grape ivy basket. (Courtesy of Marshall Horsman)

is that their native habitats are almost all over the world (which few plants are)—from Africa to Australia to the southern United States to Arabia to Southeast Asia. The most common type, *Cissus rhombifolia,* grows natively from Mexico to Colombia and Brazil as well as in the West Indies. Its name comes from the rhombus-shaped leaves.

Cissus is listed as tropical, though many species grow in the more arid areas of the tropics and also at higher altitudes, where temperatures are lower. In my opinion, a grower should consider and treat them as subtropical. *Cissus antarctica,* the kangaroo vine, is a truly subtropical native of Queensland, Australia.

USES

Grape ivies are typically produced in freestanding pots from 3 to 10 inches (7.5 to 25 cm). Hanging basket production outpaces pot production in many areas. Typical basket sizes range from 6 to 22 inches (15 to 55 cm), with many sizes in between. The plant is useful because it tolerates very low light down to 75 f.c. (0.8 klux). Grape ivies are popular as landscape vines and ground covers in California and other areas with a Mediterranean climate.

VARIETIES

Cissus rhombifolia, the standard variety, is a climbing vine with brown stems, and its rhombus-shaped leaves and stems are somewhat hairy, especially underneath. The plant has tendrils, like many vines, and its color varies from fresh green to deep metallic green. Mandaiana is a sport of *C. rhombifolia* that was introduced in 1935. More upright than *rhombifolia,* it features a compact habit and deep green, waxy foliage. The stems are also thicker. Mandaiana Compacta is also occasionally grown, and it looks like you would expect it to. A less common *C. rhombifolia* cultivar is Fionia, which has large, deeply lobed leaves on short stalks and a compact habit.

Ellen Danica, a popular cultivar, originated as a mutation of *C. jubilee* (Nielsen-Denmark), discovered in Odense, Denmark, around 1965. It has more deeply lobed foliage than either *C. jubilee* or *C. rhombifolia,* giving the leaves more of an oak-leaf appearance. *C. antarctica,* a rather vigorous vine with lighter colored foliage, is said to be less durable in the interior environment.

PROPAGATION

Cissus will grow from cuttings or seeds. Most nurseries propagate single-node cuttings. The top of the cutting is trimmed about one-fourth inch (0.6 cm) above the node, and the more basal part of the cutting is left about 1 inch (2.5 cm) below the node. You therefore have a cutting about $1^1/_4$ to $1^1/_2$ inches (3 to 3.5 cm) long; place about six cuttings per pot. Rooting and final plant quality will be better if the cuttings are spaced uniformly within the container.

Rooting hormones, such as 0.3% IBA, are generally beneficial. Mist the cuttings very sparingly, usually only a few times per day. Stock plants can be maintained in beds or hanging baskets, with baskets taking less greenhouse space.

CULTURE

Growing grape ivies in shadehouses is generally discouraged, as the grower has inadequate control of the environment. You are better off growing grape ivy in a greenhouse with a controlled environment. Fan-and-pad cooling is essential in warmer areas, as is protection from rain. The most favorable temperature range is 68 to 82F (20 to 28C). Higher temperatures cause plants to break down. Typical light levels are 1,200 to 2,000 f.c. (12.9 to 21.5 klux), with 1,500 f.c. (16 klux) being ideal.

It is best to have a fibrous, heterogeneous mix, typically made from four or five different ingredients, to provide a varied microclimate for the roots. A forgiving soil mix that is not too absorbent is preferred, so you may irrigate regularly without fear of damping-off. Many books say grow grape ivy at a pH of 5.5 to 6.2, but the lower end of the scale is preferable.

NUTRITION

In my experience grape ivies are acid-loving plants. Some of the best grape ivies I have ever seen were top-dressed with acidic, sulfur-coated, granular fertilizers. This creates an amazingly deep, rich green and abundant new growth. It is helpful to incorporate trace elements from sulfate sources into the mix. Research shows that anything more than 5.6 grams of a three-month, coated 19-6-12 fertilizer may be counterproductive. In general, try not to exceed 8 grams of slow-release fertilizer per 6-inch (15-cm) pot unless you are using a long-term formulation.

DISEASES

Almost all of the important diseases of grape ivy are of fungal origin. Anthracnose, caused by the fungus *Colletotrichum,* is common, especially in rooting beds. Water-soaked, round lesions appear, with rings of fruiting bodies (acervuli) appearing like black pepper on the undersides of the leaves. This fungus will take out cuttings in propagation when too much mist is applied. The lesions may turn tan to gray in color. Careful water management and sprays of thiophanate methyl fungicides or Zyban (thiophanate methyl plus mancozeb) help.

Gray mold, caused by *Botrytis,* can occur in more northern climates and in winter months. Large, gray blotches, possibly with fuzzy fruiting bodies, may appear. Air movement helps some, as does cleaning up any decaying plant debris. Zyban (thiophanate methyl plus mancozeb) sprays are somewhat helpful. Getting the temperature up is also effective.

Powdery mildew, caused by *Oidium*, is common on grape ivies in more tropical regions and especially in interior environments. A thin, white, powdery fungus appears on the leaves and stems, looking somewhat like the plants were sprinkled with talc. Sprays of Bayleton (triadimefon) are very effective.

Rhizoctonia solani can cause severe blight and dieback, especially in hot weather. It frequently starts as a spot and moves very quickly, displaying reddish brown webbing on the leaves and stems. It can rapidly cause severe foliar browning, frequently in less than a week. Soil drenches of Terraclor (PCNB) help the roots, and Zyban sprays (thiophanate methyl plus mancozeb) are also effective. Remove and discard badly affected plants.

Pestalotiopsis can cause a severe cutting rot, especially in Canada. Cuttings wilt fairly quickly, then die. Thiophanate methyl fungicide may be somewhat helpful.

INSECT AND MITE PESTS

Aphids are occasionally observed on newer leaves, especially in spring and summer. Sprays of Orthene (acephate) or Mavrik (tau-Fluvalinate) control them. Caterpillars and leafrollers, which cause chewing injury, can be controlled with Orthene, Dursban (chlorpyrifos), or *Bacillus thuringiensis* insecticides such as Dipel. Root and foliar mealybugs attack grape ivy. Granular Disyston (disulfoton) or sprays of Orthene help. Root mealybugs are usually treated with drenches of diazinon, but be careful of phytotoxicity.

Spider mite infestations can be easily missed in grape ivy, as the webbing of the mite can be disguised by the hairy nature of the plants. Mite-infested leaves turn yellow or speckled. Popular miticides include Mavrik (tau-Fluvalinate), Kelthane (dicofol), and Pentac (dienoclor). Broad mites, which you cannot see without magnification, cause branch tip necrosis and abortion, with extremely distorted leaves. Sprays of Kelthane or Pentac help, especially when used with a surfactant.

DISORDERS

Cissus plants are easy to overfertilize. Elevated soluble salts cause leaf tip burn, with wilting and loss of the older foliage. The plants also fade and die back when temperatures are too high. Iron deficiency causes mild chlorosis in the new foliage. Lack of nitrogen results in an overall pale plant with little growth. When sulfur is deficient, plants show pale, pea green color. Grape ivies are proven to grow best with relatively little fertilizer, regular watering, and a medium that is not overly absorbent.

TRICKS

Sulfur and acidity are critical for good color in grape ivy. Top-dressing with a light rate of sulfur-coated urea granular fertilizer does wonders for grape ivy. Cissus also does very well with air movement, so keep the fans running. Don't treat it like a tropical plant, as it is really only semitropical. Ornalin (vinclozolin) can be phytotoxic, as can Cygon (dimethoate). Keep Zyban (thiophanate methyl plus mancozeb) in the spray program on a regular basis, occasionally mixing in a registered miticide.

INTERIOR CARE

Cissus plants do well under low light in interiors. The minimum tolerable light level is 75 to 100 f.c. (0.8 to 1.1 klux). However, growth and quality are improved at 150 to 400 f.c. (1.6 to 4.3 klux).

Avoid letting the plants get too dry, as yellowing and abscission will result. Fertilize very lightly, and then only every three months or so, preferably with an acidic, soluble fertilizer.

 When fertilizer burns, it is actually pulling water out of the roots, and the plant suffers moisture stress.

REFERENCES

Bailey, L.H., and E.Z. Bailey. 1976. *Hortus Third*, 273-274. New York: Macmillan.

Blessington, T.M., and P.C. Collins. 1993. *Foliage Plants*, 66-68. Batavia, Ill.: Ball Publishing.

Chase, A.R., L.S. Osborne, and R.W. Henley. 1985. Grape Ivy. *Foliage Digest* (August): 5-8.

Poole, R.T., and C.A. Conover. 1990. Growth of *Cissus*, *Dracaena* and *Syngonium* at Different Fertilizer Levels, Irrigation Frequencies, and Soil Temperatures. *Foliage Digest* (August): 1-4.

Watkins, J.V., and T.J. Sheehan. 1975. *Florida Landscape Plants*, 247. Gainesville: The University Presses of Florida.

CODIAEUM

Croton

HABITAT

Codiaeums, commonly called crotons, have been enjoyed by Pacific islanders for centuries. Crotons are native to the Moluccan Islands, between the Philippines and New Guinea, and not from the Malay Peninsula, as is commonly reported. The plants were first formally studied by Dutch naturalist G. E. Rumphius prior to 1690. He named the plant *Codiaeum* after the Malaysian name for it, *codebo*. In 1762 Carl von Linne applied the common name "croton" after an ancient Greek city, Croton. The plants spread to England by 1804 and became quite popular. The first ornamental crotons for sale in the United States were marketed in Philadelphia as early as 1871, and some Florida nurseries began offering crotons about 1886.

Vivid, variable leaf coloration gives crotons their popularity. The yellow and green colors of the Brazilian flag are said to have originated from croton colors. The plants are tropical shrubs, usually grow-

Fig. 14 Croton Petra stock plants grown in Costa Rica. (Courtesy of Bill Lewis)

ing no more than 6 feet (1.8 m) tall. Their habitat is equatorial, with high rainfall and generally acidic soils. The plants are adapted to warm temperatures, shifting light levels, high humidity, and frequent rains.

USES

Crotons are usually produced as container specimens in 3- to 14-inch (7.5- to 35-cm) pots. Smaller containers usually have one plant per pot, whereas the larger ones may have three or four. The generally large leaf size creates a full, bushy effect even without many plants in the container. Rooted croton cuttings are occasionally used in dish gardens, as well. In landscapes crotons serve as mass plantings, hedges, accent plants, and foundation plants. They are seen planted around homes in virtually every part of the tropical world. Interestingly, croton leaves have been used as clothing and headdresses for centuries in South Pacific islands.

VARIETIES

Crotons can be crossed rather easily, and certainly over 200 varieties exist. It is said that around 1901 Philadelphia nurseryman Robert Craig had a croton collection that suffered severe cold injury one night. The plants, thought to be dead, were bunched in a corner, where they proceeded to flower and cross-pollinate. When the seeds were planted, a cluster of new varieties was produced, and the practice of croton hybridization spread.

There are at least 165 landscape varieties, with various leaf sizes and shapes, including linear, corkscrew, oak, broad, narrow, recurved, and interrupted. Coloration, of course, varies widely with variety and cultural conditions.

When used as foliage plants, however, the landscape varieties tend to drop leaves and lose color quickly. Several types have been bred to be well adapted to low light conditions. Many of these come from Europe, where tolerance to low light is critical. Popular foliage varieties of crotons include the broad-leaved Norma and Petra, which are excellent for medium to large container production. Mammey is a wildly colored, slow-growing corkscrew type. Such smaller-leaved cultivars as Gold Dust and Banana, with attractive yellow-and-green foliage, are good for small to medium containers. It is usually best to stick with these and other European-bred varieties rather than to try the garden types.

PROPAGATION

In the humid tropics croton cuttings are frequently stuck directly into the ground, where they root and flourish. Crosses are initially propagated from seed. It is important to harvest and plant the seeds as they start to lose their green color but before they turn black. In the commercial world tip cutting and air-layering are the most common propagation methods.

Stock plants generally yield between 10 and 20 cuttings per plant per year. Cuttings are frequently purchased from sources in the Caribbean, Central and South America, and Europe. Avoid sticking long basal stem pieces deeply into the soil. Cuttings from central stem pieces (joints) can be rooted but take longer to finish. Cuttings may be stuck directly in the finishing container, or set in liner trays or small pots for repotting later. Rooting hormone is helpful (0.1% IBA or similar), and it helps to make a hole in the medium with a pencil, then insert the cutting, in order to avoid removing the rooting hormone.

Cuttings are usually rooted under mist or tents. Misting about five seconds every 10 minutes in open shadehouses is common. Less mist is usually required in a closed greenhouse. Placing cuttings in a tightly closed polyethylene tent to retain 100% relative humidity is also successful, especially in combination with bottom heat. If cuttings are moisture stressed anywhere along the way, severe leaf drop can occur. Rooting is usually accomplished in three to four weeks, though it may take up to twice as long in winter.

CULTURE

Leaf coloration is the main reason crotons are grown, and color patterns are significantly affected by light levels and temperature. The plants actually grow in anywhere from 500 to 15,000 f.c. (5.4 to 161 klux). The good European hybrids perform well between 2,500 and 3,200 f.c. (27 to 34.5 klux), and plants will be well acclimated to interior conditions when grown at these levels. However, many growers prefer to grow crotons under higher light levels for the better coloration and faster crop turn. Shade from 55 to 63% is common in tropical shadehouses. Most varieties tolerate between 40 and 100F (4.5 and 38C), with the preferred temperature range being 65 to 90F (18.5 to 32C).

Crotons are not very particular about potting media, but they need at least reasonable drainage to have decent root systems. They tolerate frequent irrigation if drainage is good. Color is enhanced under higher light conditions and with higher potassium fertilization. High-nitrogen fertilization tends to reduce leaf color. Target pH is usually around 5.5 to 6. One cutting per pot is typical for smaller containers, three to four cuttings for larger containers. Spread cuttings well apart in the pot.

NUTRITION

Crotons are most frequently grown with a 1-1-1 ratio of N-P_2O_5-K_2O. Constant liquid feed at 200 ppm N is common for small to medium containers. Growers producing larger specimens frequently use a coated triple 13 or triple 14 or a granular fertilizer as a topdress. Trace element problems are rare in crotons, but you should have some form of minor elements in your fertilizer program. Sulfur is an important nutrient for croton color contrast. Lack of sulfur may cause a certain subtle dullness in coloration. Sprays of magnesium sulfate help generate more vivid, contrasting color.

A useful technique is to push crotons with nitrogen early in production to achieve plant size, then let fertility decrease as plants mature, in order to have finished plants with bright, contrasting color patterns. Excessive fertility, though, causes bleaching of foliage and reduced color contrast, even when light is ample. In northern climates liquid fertilization with calcium and potassium nitrate is popular for maintaining strong plants in low light conditions.

DISEASES

The most commonly encountered croton disease is anthracnose, caused by the fungus *Colletotrichum*. *Colletotrichum* leaf spot is a major problem on cutting farms in wet locations. It may start as a small water-soaked leaf spot anywhere on the plant, and the mature lesion frequently becomes red and fairly circular. Fruiting bodies that look like black pepper may appear on the underside of the lesion. Keeping foliage dry is very helpful, as are sprays of Dithane (mancozeb) or a thiophanate methyl fungicide.

Keep propagation mist heads absolutely level for even distribution of mist.

Bacterial crown gall, caused by *Agrobacterium tumefaciens,* causes an odd swelling of stem tissue. There is no chemical control, and growers have learned to recognize and destroy plants infected with crown gall. Because of this, crown gall is rather rare today. A similar fungal crown gall is caused by *Kutilakesa*. This fungal crown gall can be controlled somewhat with thiophanate methyl sprays. Growers encountering crown gall should have the disease diagnosed by a plant pathologist to determine whether to dump the plants or to treat them.

Rhizoctonia can cause severe foliar blight, especially when plants are closely spaced in propagation. A quickly spreading, dark brown to black rot will cause severe defoliation and plant death. Sprays of Chipco 26019 (iprodione) can help prevent it. *Pythium* root rot occurs when the medium stays too wet. Careful watering, a well-drained mix, and drenches with Subdue (meta-laxyl) or Aliette (fosetyl-aluminum) should control it.

Insect and mite pests

Spider mites are the major pest problem of crotons. Both two-spotted and red spider mites are frequently encountered. Because of the thickness of croton leaves, mites primarily attack immature foliage. As the mite-damaged leaves grow out, they become misshapen, with broad lobes or serrations along the leaf edges. The grower generally needs to spray a miticide preventively. Common choices include Pentac (dienochlor), Avid (abamectin), and Thiodan (endosulfan).

Mealybugs are an occasionally severe pest of crotons. The white, cottony masses are generally seen on the undersides of leaves and on the petioles near the stem. Several sprays with either Diazinon, Cygon (dimethoate), or Talstar (bifenthrin) should take care of the problem. Several scale insects attack crotons. They are frequently found either on stems or on the undersides of leaves. Chemical controls are similar to those for mealybug.

Thrips frequently cause a scaly- or raspy-looking injury, frequently observed as a longitudinal streak on the leaf underside. Sprays of Thiodan (endosulfan), Mavrik (tau-Fluvalinate), or several others are helpful. Concentrate the spray near the growing tips, where the thrips are active.

Disorders

Elevated soluble salts and high nutrient levels cause bleaching and discoloration of the foliage. Leaching and reducing fertility can help color return, though normally only to subsequent leaves. Crotons do not grow if planted too deeply. This can cause significant irregularity of cutting size in a finished 8- or 10-inch (20- or 25-cm) container.

Crotons change their color patterns somewhat when exposed to temperatures in the 40s F (5 to 8C). The Norma variety, especially, turns quite red. Severe cold causes defoliation. In cutting farms sections of the stock can become dull in color. This may be due in part to genetic reversion and age. It is helpful to cut out the reverted stock and avoid propagation of dull-colored cuttings.

Tricks

Sprays or liquid fertilization with potassium nitrate help improve color intensity in most crotons, regardless of light level. The plants send out longer roots if grown with bottom heat. Old, tired-looking crotons frequently respond to a topdress of dolomitic lime at the rate of 1 tablespoon per 6-inch (15-cm) pot. Malathion may be phytotoxic on some varieties, so I avoid using it in the spray program. Crotons respond very well to leaf shine products prior to shipping. The leaf shine also tends to reduce foliar residues from spray or hard water, which would diminish the plants' attractiveness.

Growing stock plants in smaller-than-normal containers gives you more compact cuttings. With some varieties, removing only about two leaves from the base of the stem and sticking the cutting very shallowly gives faster rooting and ultimately a better plant. The rest of the leaves on the cutting serve to hold the cutting vertical.

Croton stem juice stains clothing, looking like blood stains when dried. Dipping in a mixture of equal parts chlorine bleach, borax, and water may remove the stains, in some cases.

INTERIOR CARE

Crotons need somewhat more fertilizer and more light than most interior plants. Every six weeks or so, an application of soluble 24-8-16 or 20-20-20 can be helpful, but keep the rates conservative. Color is maintained if the plants receive 500 to 1,000 f.c. (5.4 to 10.8 klux), though some cultivars tolerate less light. Irrigation requirements are about average. Avoid exposing the plants to cold drafts, or rapid leaf drop may occur. Keep a sharp eye out for spider mite activity, and spray with insecticidal soap or a similar product if mites are observed.

REFERENCES

Brown, B.F. 1960. *Florida's Beautiful Crotons.* Eau Gallie, Fla.: Undersea Press.

Conover, C.A., and R.T. Poole. 1983. *Cultural Factors Affecting Croton Propagation.* ARC-A Research Report RH-83-21. Apopka: Univ. of Florida Agricultural Research Center.

Dekker, J. 1982. Tips on Growing Crotons. *Florist's Review* (July 22): 34.

Langefeld, J. 1985. Culture notes. Codiaeum variegatum pictum (Croton), Family: Euphorbiaceae. *GrowerTalks* (January): 18-22.

Lukas, C. 1984. Crotons of Many Colors. *Interior Landscape Industry* (September): 25-28.

Osborne, L.S., A.R. Chase, and C.A. Conover. 1984. Croton. *Nurserymen's Digest* (June): 44-46.

Reeves, R.G., and A. Bell. 1988. *Codiaeum variegatum pictum:* A Foliage Favorite With New Possibilities. *Florida Nurseryman* (January): 70-75.

COPPER LEAF PLANT

(See Other Foliage: Acalyphas*)*

CORDYLINE

Ti Plant

HABITAT

Cordylines are somewhat similar to dracaenas in that, both in the Agave family, they share several characteristics. The common name for cordyline is in fact "dracaena," though plants truly in the genus *Cordyline* have five to 16 ovules per cell in the flowers, whereas those in the genus *Dracaena* have only one.

There are about 20 species of *Cordyline*, most native to India and Australia. The name comes from the Greek word for "club," referring to the plant's thickened roots. *C. terminalis*, the name of the principal cultivated species, refers to the end of the

Fig. 15 Cordyline plants growing at a stock farm in El Salvador. (Photo by the author)

92

stem. *C. terminalis* is native to the moist tropical forests of eastern Asia, where it attains a typical maximum height of 10 feet (3 m). Cordylines are strictly tropical, adapted to warm, humid areas with abundant rainfall. Being understory forest plants, they prefer shaded conditions, though they tolerate some direct sunlight. Exposure to windy conditions results in physical damage to the foliage.

USES

Commonly called ti plants, the good luck plants, or Hawaiian ti, cordylines are attractive and durable both as foliage and as landscape plants. The cultivars can include purple, maroon, rose, pink, or yellow in addition to green. The highly variegated foliage gives them a somewhat painted appearance. Smaller varieties, such as Baby Doll, probably the most widely grown cordyline, are grown for small containers and dish gardens. Numerous larger types are produced commercially as houseplants and interior accent plants. Typical container sizes range from 6 to 17 inches (15 to 42.5 cm).

The long, tapered, colorful leaves are distinctly tropical in appearance, and the plants' upright habit makes them desirable as medium-sized interior specimens. Cordylines are also extremely popular as landscape plants in the tropics, especially in the South Pacific, Caribbean, and Central American regions, usually in the lower altitudes.

VARIETIES

The basic *Cordyline terminalis* is a solid green, nonvariegated plant. The cultivated varieties all have variegation to some degree. Baby Doll is a dwarf, somewhat narrow-leaved variety with maroon, pink, and green foliage. Red Sister is larger, with attractive contrasting foliage of red, rose, and pink. Kiwi, smaller than Red Sister, has red- and green-striped foliage, presenting a somewhat painted or Southwestern-style appearance.

There are a number of larger cultivars with various color schemes, although the names are not consistently followed. Useful varieties include Black Star, Snow King, Tri Color, Bolero, Eugene Andre, and Firebrand. Some very attractive but lesser known cordyline cultivars are coming out of Holland and Central America. The discriminating cordyline grower should seek out some of these varieties, as they are colorful and attractive houseplants with very little competition in the marketplace.

PROPAGATION

Cordylines can be produced from seed, cuttings, root layers, and air layers. Tip cutting is the most common type of commercial propagation, though air layers are sometimes grown for the larger cultivars. Numerous field-grown stock farms exist in the Caribbean and Central American regions as well as the Pacific Islands. Baby Doll is one of the more prolific producers, generating about 50 cuttings per square foot per year (535 per m²). The larger cultivars take up somewhat more space, yielding lower but still acceptable production rates.

Cordyline tip cuttings, usually purchased unrooted, are stuck in the finishing container after removal of two or three basal leaves. Air layers are transplanted similarly, though with more care to avoid injuring the roots. Mist is occasionally applied manually, typically two or three times per day. Unrooted cuttings can take up large amounts of fluoride in a short time, so mist and irrigation should be low in fluoride to avoid excessive leaf loss and tip burn. Antitranspirant sprays are sometimes used to retard necrosis and loss of older foliage. Bottom heat is used for propagating media in colder climates.

CULTURE

Cordylines are valued primarily for their multicolored, variegated foliage. Like with crotons, the variegation is affected by temperature, light, and fertility interactions. The cordylines are usually grown at 3,000 to 3,500 f.c. (32.3 to 37.7 klux), or about 63 to 73% shade, in the tropics. The ideal temperature range is 65 to 95F (18.5 to 35C). In many cases you are actually shading for temperature reduction as much as for light reduction.

The potting medium should be well drained, with no more than 50 to 60% of high-quality peat as a substrate. The medium should be well limed to around 6.5 pH and have good levels of calcium. Superphosphate should not be incorporated into the medium, and try to use low-fluoride phosphate sources. Irrigation water should preferably have less than 0.2 ppm fluoride, at most no more than 0.25 ppm.

NUTRITION

Moderate amounts of 3-1-2 fertilizers are generally used. Coated, slow-release fertilizers are popular, as their release rates coincide with the moderate rate of plant growth. Such fertilizers, which are generally low in fluorides, can be top-dressed or incorporated into the medium at the middle rate. It is essential to have good calcium nutrition and low-fluoride fertilizer sources for cordylines.

Another common nutritional program is liquid feed with 24-8-16 or 20-10-20. These are generally low-fluoride formulations, as well. Liquid feed can be applied constantly at about 150 ppm, or one to two times per week at 300 ppm nitrogen. Liquid fertilizer should be rinsed off of the foliage by letting clear water run briefly after the feeding. This helps remove fertilizer salts from the terminal whorl and avoids death of the emerging leaf.

Ti plants have rather high magnesium requirements, and lack of magnesium or potassium can greatly affect leaf variegation patterns. Deficiency generally shows up in older foliage, especially in plants that have been cut back. Lack of magnesium shows as broad yellowing in the older leaves, especially toward the tips. Plants lacking potassium have very dull color in the older foliage, frequently with small, necrotic spots.

DISEASES

Perhaps the most serious disease of cordylines is bacterial soft rot. It can be caused by two different types of *Erwinia, E. chrysanthemi,* and *E. carotovora* pv. *carotovora. Erwinia* causes a wet, slimy, smelly rot of leaves, stems, or roots. Cuttings in propagation can be affected severely, tending to rot from the bases upward. Chemicals may help preventively, but they may not do much once disease symptoms have begun. Discard affected plants and media, then spray the rest with Phyton 27 (PCAF) or another registered copper fungicide (Agrimycin 17 is not labeled for cordylines, to my knowledge).

The fungus *Fusarium* can also cause a dry stem rot, the bark becoming loose and lesions observed near the base of the stem. It differs from *Erwinia* in that the internal bark tissue is generally dry and lacking odor, though the roots may be water-soaked, as with *Erwinia.* Keeping the plants off the ground will help, as will keeping the soil pH up and using thiophanate methyl fungicides, such as Cleary's 3336.

Several leaf spots attack cordylines, including *Fusarium moniliforme,* which causes oval, reddish brown spots with yellow halos on the newer leaves. The spots may be somewhat orange in appearance. Keeping foliage dry is helpful. Sprays of Daconil (chlorothalonil) or Chipco 26019 (iprodione) will control it if they are labeled for your situation. *Phyllosticta* causes circular spots on older leaves. The spots are tan with a purple border and a yellow halo. The entire leaf will become necrotic if disease pressure is heavy. Dithane (mancozeb) is effective, along with keeping the foliage dry. *Phytophthora* causes occasional leaf blight under wet conditions. Drenches with Subdue (metalaxyl) or sprays of Aliette (fosetyl-aluminum) help. Definitely reduce soil moisture when fighting *Phytophthora*.

Finally, southern blight attacks cordylines under hot weather conditions and under the heat of tents. Large amounts of white mycelial growth can be seen on soil and plant surfaces, as can little white or tan sclerotia, which look like mustard seeds. Terraclor (PCNB) is a useful fungicide, but if you need to repeat treatment in the same area, the insecticide Dursban (chloropyrifos) is pretty good. Southern blight is usually not a problem in cooler situations.

INSECT AND MITE PESTS

Two-spotted mites are by far the major pest problem for cordyline growers. Mite feeding causes speckles on the foliage, reduced color, and in severe cases leaf loss and webbing. Cordyline growers usually spray preventively with miticides, such as Kelthane (dicofol), Mavrik (tau-Fluvalinate), Pentac (dienoclor), or Avid (abamectin), or with insecticidal soaps. Predator mite programs are also useful.

Leafhoppers occasionally cause injury on older leaves. Orthene (acephate) sprays should be effective. Mealybugs are also fairly common, appearing as cottony masses on the leaf axils. Sprays of Cygon (dimethoate) or Dycarb (bendiocarb) will be helpful, although several applications may be needed. Granular Di-syston (disulfoton) also controls mealybugs. Scale insects occur from time to time, especially Florida red scale (*Chrysonphalus aonidum*). They usually come in on the cuttings as a very light, undetected infestation. The scale insects will proliferate substantially by the time they are noticed. Florida red scale looks like a small, reddish brown disk with a lighter colored spot in the center. Treat using mealybug control measures.

Thrips cause silver to gray leaf scars, usually on the leaf undersides. Sprays of Mavrik (tau-Fluvalinate), Dycarb (bendiocarb), or Orthene (acephate) should be helpful. Fungus gnats may be observed as small, black flies buzzing around the soil surface. Their larvae can feed on roots. Fungus gnats are primarily a problem in wet conditions and when large amounts of algae are present. Controls include drying down the greenhouse and treating with Diazinon or Gnatrol (*Bacillus thuringiensis*).

Disorders

Baby Doll cordyline was one of the first plants discovered to be sensitive to fluoride, and the cordylines remain among the most fluoride sensitive of foliage plants. Severe tip burn, escalating into full-blown marginal necrosis, is frequently encounted with fluoride toxicity. The symptoms look very much like a blight disease. Keep soil pH and calcium up, and do your best to use low-fluoride fertilizers and irrigation water.

Weak color in cordylines is usually caused by problems with temperature, light, or fertilizer. High temperatures cause dull color. Cordylines frequently have better color between November and May in the Northern Hemisphere because of cooler night temperatures. Light levels above or below the recommended range tend to cause color abnormalities. Excessive fertilizing or imbalanced fertilizing, such as lack of magnesium, also reduces color.

Losing the emerging leaf at the growing tip, due to failure to rinse liquid fertilizers, is quite common, as cordyline foliage is very sensitive to salts. Be sure to rinse liquid feed off the foliage, and don't spray with high rates of soluble fertilizers.

 Fluoride-sensitive plant varieties almost always have long, tapered leaves.

Tricks

Cordylines prefer a tall greenhouse, preferably at least 12 feet (3.65 m) or more. A taller greenhouse helps dissipate excessive heat during hot summer conditions, and the temperature reduction helps maintain color and reduce fluoride symptoms. As you grow cordylines at higher altitudes, you can grow them with steadily less shade. Heat and light are once again interrelated, and much of the shading is done to reduce temperature rather than light levels. Also, during warm periods irrigate in the middle of the day to increase humidity and reduce leaf temperature.

If the basal part of cuttings begins to rot, you can pull up those cuttings and recut them to healthy tissue with sterilized clippers. You may still lose some of the cuttings, but you will save some. Once cuttings begin to rot, they will all be lost unless this technique is followed.

Do not irrigate with fluoridated tap water, nor with high-fluoride irrigation water. Keeping soil calcium levels up helps bind fluorides. (Your teeth are largely made of calcium fluoride.) Calcined clay in the mix is one way of doing that. Also, check your liming sources for fluoride because many can contain significant amounts. Leach perlite prior to incorporating it into the medium in order to help remove fluorides. Avoid heat stress on cordylines exposed to fluoride. As the fluorides accumulate in the leaf tips, they prevent the closure of the stomates; hence, a localized burning ensues.

INTERIOR CARE

Ti plants are rather well adapted to low-light interior situations. The plants can handle 75 to 150 f.c. (0.8 to 1.6 klux). The color tends to improve, however, as the light level is increased. Under low light, new growth is stretched and poor in color. If you must keep the plants under very low light, avoid fertilizing them.

Cordylines like humidity, preferably 40 to 60%. Try to keep temperatures between 65 and 85F (18 and 29C), and avoid cold, drafty areas. Be careful with leaf shine products, as some cordyline varieties are sensitive to some brands.

The primary problems encountered by cordylines in the interior are fluoride injury and spider mite infestations. Do not use fluoridated tap water on these plants. Collected rainwater is probably the best water source, followed by low-fluoride well water. Plants should be kept evenly moist, avoiding extremes in either direction.

REFERENCES

Bailey, L.H., and E.Z. Bailey. 1976. *Hortus Third*, 312, 398. New York: Macmillan.

Blessington, T.M., and P.C. Collins. 1993. *Foliage Plants: Prolonging Quality*, 75-78. Batavia, Ill.: Ball Publishing.

Conover, C.A., and R.T. Poole. 1972. Production of High Quality Plants of *Cordyline terminalis* 'Baby Doll'. *Florida Foliage Grower* 9 (8): 1-2.

Osborne, L.S., C.A. Conover, and A.R. Chase. 1985. Cordyline (Ti Plant). *Nurserymen's Digest* (August): 74-77.

Watkins, J.V., and T.J. Sheehan. 1975. *Florida Landscape Plants*, 108. Gainesville: The University Presses of Florida.

CORN PLANT

(*See* Dracaena fragrans)

CROTON

(*See* Codiaeum)

DATE PALM

(*See* Phoenix)

DIEFFENBACHIA
Dumb Cane

HABITAT

Dieffenbachias are among the most familiar foliage plants around the world. They are commonly known as dumb canes in many areas because oral exposure to the plant's sap results in a swelling of the tongue and an inability to speak. Named after a German physician, Dr. J. Dieffenbach, the genus has about 30 species that are native to tropical America from Costa Rica to Colombia. Over 100 cultivated varieties exist, as dieffenbachias hybridize fairly easily. Most of the varieties in the foliage trade come from *D. amoena* and *D. maculata.*

Fig. 16 Four dieffenbachia varieties growing in a Jamaican nursery. (Photo by the author)

In the wild, dieffenbachias generally grow in low, moist areas. They are frequently found as understory plants in moist, lowland tropical forests and therefore tend to prefer somewhat shaded and moist locations. Dumb canes are frequently observed fairly close to the sea, and they do have some salt tolerance. The plants can also be found growing wild on the banks of the Amazon River.

Uses

Dieffenbachias are generally desired for their multicolored foliage. The leaves are broad, flat, and somewhat brittle. The plant is commonly seen as a houseplant, office plant, or in interior planters for a tropical effect. Growers commonly produce dieffenbachias in pot sizes ranging from 3 to 10 inches (7.5 to 25 cm). Smaller containers are usually grown with one plant per pot. Larger varieties may be grown one plant per pot if they sucker well, though sometimes three cuttings per pot are used for varieties that tend not to sucker, such as Tropic Snow. The smaller *D. maculata* varieties are frequently seen in dish garden combinations.

Varieties

While more than 100 cultivated varieties exist, the majority of foliage production is limited to about one dozen cultivars. One larger type is *D. amoena* cv. Exotica, which has deep green foliage with irregular white blotches. A mutant of *D. amoena* called Tropic Snow has a bolder, more attractive variegation pattern than Exotica. This plant was discovered at Chaplin's Nursery in Fort Lauderdale, Florida, and has been in the trade for many years.

Two of the smaller *D. maculata* cultivars include Perfection Compacta, a dwarf selection from *D. maculata* Perfection that produces many basal suckers. Marianne, developed as a mutant of Perfection, has bright yellow leaves with dark green margins. Camille, a more popular variant of Marianne, has similar leaf coloration. The Maui looks somewhat like a larger version of Camille and tends to be grown in lower light. Rudolph Roehrs is an older, medium-sized variety with yellow to chartreuse foliage. Wilson's Delight is a larger, sturdy variety with deep green foliage offset by a bright white midrib. The sap from Wilson's Delight, unfortunately, can be very irritating to the skin. Other new patented varieties in the trade include Starbright, Starwhite, and Sparkles.

PROPAGATION

Depending on variety, almost all commercial propagation of dieffenbachias is done by either tip cuttings or tissue culture. The larger Tropic Snow types tend to be direct-stuck in the finishing container. Cuttings are best taken with sterilized knives; cut the stem at an angle for more surface area, but do not cut through the node. Good stem caliper of cuttings is important for an attractive finished plant. Cuttings typically lose one or two leaves during the rooting process, which takes two to three weeks. Rooting hormones are sometimes used, but they are not considered essential for this plant. Cuttings are generally rooted without mist or perhaps with one brief misting during the day. Excessive misting, though, leads to severe disease problems.

When rooting smaller varieties, such as Camille and Compacta, from cuttings, it helps to have a fairly small cutting, generally no more than 3 inches (7.5 cm) long, with good stem caliper. A cutting of this type results in a plant whose basal shoots are in proportion to the main stem. A longer cutting would result in a tall principal shoot, with smaller suckers at the base. Such a plant is generally not marketable.

Dumb canes can also be propagated from leafless stem pieces rested horizontally in the soil, or from ground layering. These techniques, however, are better suited to hobbyists and collectors than commercial growers.

Much dieffenbachia propagation over the years has shifted to tissue culture, partly in an effort to avoid diseases such as *Erwinia* and dasheen mosaic virus. Many of the newer cultivars are produced almost exclusively via tissue culture. Well-grown tissue-culture liners also tend to produce plants with better uniformity and pot habit.

Culture

Dieffenbachias are generally grown in covered greenhouses at light levels ranging from 1,500 f.c. (16.2 klux) to 3,000 f.c. (32.3 klux). They can be grown in shadehouses, but at times excessive rains cause disease problems. They can be grown on ground cover, but most growers prefer to grow them either on benches or at least on inverted pots. If the light levels are too low, the plants will look good but will grow very slowly. If the light levels are too high, the foliage tends to be very erect, with petioles pointing almost straight up. Foliage may be bleached, perhaps with marginal burn, and smaller than usual. Most growers produce the plant in 75 to 80% shade, depending on variety, time of year, and other factors.

The critical low temperature is generally considered to be 50F (10C), though some varieties will take 45F (7.2C) without injury. The plant tolerates occasional cool nights, but too many cool nights significantly slow plant growth. For best results, keep night temperatures up on a consistent basis.

Dumb canes tend to do better in lightweight potting media formulated with sphagnum peat, rock wool, or coconut fiber as a principal ingredient. Heavier mixes of the sedge peat type tend to be associated with dieffenbachia disease problems, unless the plants are irrigated very carefully. Many growers like to add perlite, styrofoam beads, or wood products to the medium in order to have an open mix that doesn't retain excess moisture.

Overwatering is a critical problem in dieffenbachia production. The medium needs to dry out fairly well prior to irrigating, and water should be applied when drying conditions are good so as to minimize the amount of time the foliage stays wet, or else disease pressure increases. If plants are allowed to go a little too dry, the older leaves start to turn a solid canary yellow.

The suckering varieties tend not to send out as many basal shoots if plants are grown close together. Some growers place the plants pot to pot for a few weeks to save space, then place them on a wider spacing once they begin to grow. Dumb cane needs some degree of light coming in from the side in order to sucker well and for symmetry.

Nutrition

The nutrient requirements of dieffenbachias are about average for foliage plants. A 3-1-2 ratio of N-P_2O_5-K_2O is preferred, at least for soilless media. Trace elements should not be overlooked, as the plant tends to respond fairly

TABLE 9 Leaf analysis rating standards for *Dieffenbachia maculata*

Nutrient (%)	Very low	Low	Medium	High	Very high
Nitrogen	<3.0	3.0–3.2	3.3–5.0	5.1–6.5	>6.5
Sulfur	<0.20	0.20–0.24	0.25–0.50	0.51–0.75	>0.75
Phosphorus	<0.20	0.20–0.24	0.25–0.75	0.76–1.20	>1.20
Potassium	<2.00	2.00–2.45	2.50–5.50	5.55–7.00	>7.00
Magnesium	<0.20	0.20–0.24	0.25–0.75	0.76–1.00	>1.00
Calcium	<0.80	0.80–0.99	1.00–2.50	2.51–3.00	>3.00
Sodium			<0.20	0.21–0.50	>0.50
(ppm)					
Iron	<50	50 –59	60–300	301–500	>500
Aluminum			<250	250–2000	>2000
Manganese	<35	35–49	50–300	301–500	>500
Boron	<15	15–19	20–50	51–75	>75
Copper	<6	6–7	8–50	51–200	>200
Zinc	<15	15–19	20–201	201–400	>400

Sources: Institute of Food and Agricultural Sciences, Apopka, Florida; Dr. Benjamin Wolf, Fort Lauderdale, Florida.

Notes: Common names include dieffenbachia, Exotic, Rudolph Roehrs, Tropic Snow. Sample of most recently fully mature leaves, no petioles.

well to them. Many growers use constant liquid fertilization, with a leaching every two weeks or so. Others use a combination of liquid and granular fertilizer. An occasional leach is important because growing the plants correctly on the dry side is quite difficult if soluble salts are elevated.

Chloride is an essential plant nutrient.

Because of its succulent nature and tendency to grow near the sea, dieffenbachias frequently accumulate sodium as an electrolyte, even in a medium low in salt. Sodium accumulation is not usually a concern in this plant. Potassium requirements are a little higher than in many foliage varieties. Lack of potassium tends to appear as necrotic lesions in the older foliage. The varieties with more chlorophyll tend to suffer from magnesium deficiency, especially in stock plants. Symptoms include a progressive yellowing of the older foliage from the leaf margins inward. When dieffenbachias lack nitrogen, the leaves are smaller than normal, with reduced rate of growth and lack of color contrast. Iron deficiency shows up as the typical veinal chlorosis in the new leaves. Iron toxicity looks very much like magnesium deficiency in this plant. Lack of boron will appear as longitudinal ribbing in the new leaves–without adequate boron the leaves have some difficulty unfurling. Deficiencies of manganese, zinc, and copper are fairly rare in dieffenbachias because the plants are frequently sprayed with fungicides containing those elements.

DISEASES

Numerous disease problems afflict dieffenbachias when cultural conditions are less than ideal. The most severe disease of this plant is *Erwinia*, which causes bacterial blight, leaf spot, and stem rot. Infected tissue is generally very wet and water-soaked in appearance. A mushy rot ensues, usually with a foul odor. Early symptoms may show up as premature yellowing of older leaves. The bacterium can be systemic within stems, a critical danger to stock plant growers. The disease appears to be less active during winter months. The best controls are cultural, using good sanitation measures, cutting with sterile tools, and discarding affected plants. Phyton 27 (picro cupric ammonium formate) is helpful as a chemical control. Streptomycin sulfate or mixtures of Dithane and copper help a little, but not very much.

A second major disease of dumb cane is *Fusarium* stem rot. The infection looks somewhat similar to *Erwinia* stem rot, though the edge of the infected area frequently appears purplish or reddish. You may see bright red fruiting bodies at the base of the stem. *Fusarium* can also cause a roundish, papery-looking leaf spot. In addition to sanitation measures and cultural controls, it generally helps to keep media pH up. Soil drenches or dips with a thiophanate methyl fungicide are helpful.

Dasheen mosaic virus is not uncommon, especially in plants propagated from cuttings. Symptoms include a streaking in the color pattern, occasional ring spots, mild leaf distortion, and reduced plant vigor. The virus is transmitted by cutting tools and occasionally by aphids. Symptoms tend to be more pronounced in winter. Tropic Snow frequently has latent infection from the virus, but this rarely results in a production problem. The only controls are to maintain pathogen-free stock and to keep aphids under control.

A few other diseases are also commonly seen on dieffenbachias. *Colletotrichum* causes a somewhat irregular fungal leaf spot, frequently with small, black fruiting bodies on the underside of the leaf. Sprays of Dithane (mancozeb) are generally helpful. Infected leaves should be removed. *Myrothecium* causes a more regularly shaped, large, round leaf spot, especially after leaves have been physically injured. This disease is common on newly arrived cuttings. Chipco 26019 (iprodione) is helpful as a foliar spray, as is Terraguard (triflumizole). *Phytophthora* can cause a basal stem rot under wet conditions. Subdue (metalaxyl), Truban (etridiazol), or Aliette (fosetyl-aluminum) applications are helpful, as is moisture control. *Xanthomonas campestris* pv. *dieffenbachiae* causes a fairly small bacterial leaf spot without fruiting bodies. Controls are similar to those for *Erwinia*.

INSECT AND MITE PESTS

Spider mites are by far the major pest problem for dieffenbachia growers. Two-spotted mites cause numerous small, white specks in the foliage and, in severe cases, foliage bleaching with marginal burn. Sprays of insecticidal soaps, Pentac (dienoclor), Avid (abamectin), and several other miticides can be effective. Wetting agents help in controlling mites on this plant. Morestan (chinomethionate) and Volck oil are reported to cause phytotoxicity on dieffenbachias.

Aphids are only an occasional problem on dieffenbachias, but the grower should be alert, as aphids can transmit dasheen mosaic virus. A small, reddish aphid called the rice root aphid is sometimes observed attacking the root system. Numerous chemicals for aphids are available, including Orthene (acephate). Diazinon is helpful against root aphids. Mealybugs are occasional pests of dumb canes. Dycarb (bendiocarb), Cygon (Dimethoate), and Orthene have been used successfully. Scale insects and thrips are occasional problems, though they are fairly rare on this plant.

DISORDERS

Several disorders affect dieffenbachias. A common problem is marginal burn, which can be caused by elevated soluble salts, excessive heat, or low relative humidity. Overwatering, as indicated earlier, contributes to numerous disease problems. *Erwinia* can cause water-soaked foliar lesions, but a similar symptom can occur when the soil is cool and the air is warm, resulting in breakdown of leaf tissue. When dieffenbachias are moisture stressed, one or more of the older leaves can turn solid canary yellow, and the leaf ultimately withers. Guttation, leaf tips exuding moisture in significant quantities, is observed at times. This is a physiological disorder wherein plants are taking in more water than they can transpire. Tissues showing guttation tend to become necrotic. Exposure to ethylene at a minimum of 5 ppm for three days causes premature yellowing and loss of older foliage. Ethylene produced by improperly burning greenhouse heaters may cause foliar distortion. Subdue toxicity has been observed several times in dumb canes, showing up as a yellow to white marginal band in the older leaves. Dieffenbachias are known to be sensitive to mercury in some paints. Plants exposed to mercury in the atmosphere slowly decline and lose foliage.

TRICKS

The best dieffenbachia growers select a mix that does not hold excess moisture. They then let the soil dry almost to the bottom of the pot before irrigating thoroughly, then letting the mix dry once again.

When taking cuttings, keep several sharp knives in a cup filled with bleach water, alcohol, or a disinfectant. Select a different sterilized knife each time you cut. Take the cutting at an angle between the nodes on plants with good stem caliper. Some *Dieffenbachia* varieties sucker better if the cuttings or tissue-culture liners are planted deeply.

When growing dieffenbachias as stock plants, make sure you have magnesium in the liquid feed program on a regular basis, and watch for pH drop of the medium over time. Benzyladenine sprayed at 1.2 grams per gallon (3.8 l) helps plants retain foliage during long shipping intervals. Three consecutive daily sprays of benzylaedenine at 500 ppm help some varieties send out more basal shoots.

INTERIOR CARE

Dieffenbachias do rather well in interior situations as long as they get reasonable light and are not overwatered. They need at least 150 to 250 f.c. (1.6 to 2.7 klux). They can tolerate less, but tend to stretch rather severely with time. Avoid cold, drafty situations, and try to maintain a minimum temperature of at least 65F (18C). They generally do not require fertilizer for at least several months after purchase. After about four months dumb cane will require an occasional fertilization, usually with a soluble 20-20-20 fertilizer at about 2 teaspoons per gallon (3.8 l).

Dieffenbachias should be checked periodically for spider mites, especially within the first two months after the plant is installed in an interior situation. Avoid getting the foliage wet when watering. If relative humidity drops below 25%, leaf tip burn may occur. In a low-humidity situation, it may help to set the container on a pan of moistened gravel to keep humidity up around the plant.

With time, if light and fertility are too low, leaf color and contrast will fade. If a plant eventually becomes tall and leggy, it can be cut back. The cutting can be rooted easily to start a new plant, and the original plant will branch.

REFERENCES

Ben-Jaacov, J., R.T. Poole, and C.A. Conover. 1990. Long-Term Dark Storage of *Dieffenbachia* Sprayed with Cytokinin. *Foliage Digest* 13 (3): 1-2.

Blessington, T.M., and P.C. Collins. 1993. *Foliage Plants: Prolonging Quality*, 84-90. Batavia, Ill.: Ball Publishing.

Chase, A.R. 1992. Common Diseases of *Dieffenbachia*. *Southern Nursery Digest* (December): 20-21.

Chase, A.R., R.T. Poole, and L.S. Osborne. 1983. *Dieffenbachias*. ARC-A Foliage Plant Research Note RH-1983-B. Apopka: Agricultural Research Center.

Dieffenbachia Offer Natural Differences in Size, Appearance. 1986. *Greenhouse Manager* (April): 25.

Henny, R.J. 1986. Increasing Basal Shoot Production in a Nonbranching *Dieffenbachia* Hybrid with BA. *Hortscience* 21 (6): 1386-1388.

Joiner, J.N., R.T. Poole, C.R. Johnson, and C. Ramcharam. 1978. Effects of Ancymidol and N, P, K on Growth and Appearance of *Dieffenbachia maculata* 'Baraquiniana'. *Hortscience* 13 (2): 182-183.

Rathmell, J.K. n.d. *Dieffenbachia*. Northeast Home Horticultural Fact Sheet NE-55. University Park, Penn.: The Cooperative Extension Services of the Northeastern States.

DIZYGOTHECA
False Aralia

HABITAT

Commonly known as false aralia, *Dizygotheca elegantissima* has been in the foliage trade for many years. The genus name *Dizygotheca* means "double receptacle," for the four-celled anthers on the flowers. The species name *elegantissima* refers to the rather elegant appearance of the plant, whose highly serrated or toothed leaflets feature speckled petioles and stems.

There are about 15 species of *Dizygotheca,* which are generally native to Polynesia and the South Pacific. False aralia is native to New Caledonia, east of Australia, where it grows as a small shrub or branched tree, occasionally reaching a height of 15 feet (4.6 m). The foliage is brownish, lighter underneath, with leaves being made up of seven to 11 leaflets. False aralias have raised pores (lenticels) on the stems, as do true aralias. New Caledonia is tropical but near to the Tropic of Capricorn, so false aralia will tolerate some degree of cool weather.

USES

False aralias are generally grown as free-standing specimens, with multiple seedlings per pot. Because the plant branches poorly, growers frequently plant 20 to 25 seedlings per pot to give an adequately full appearance. The unusual leaf color and texture distinguish

Fig. 17 Shade-house false aralias in Florida. (Photo by the author)

false aralia from many foliage plants. The most common pot sizes are 6, 10, and 14 inches (15, 25, and 35 cm), though larger and smaller sizes are occasionally produced. As the plant approaches about 5 feet (1.5 m), the more mature foliage pattern appears, but most aralias are grown for their more slender juvenile foliage. False aralias are also used occasionally in the landscape as small trees, usually in USDA Zone 10.

VARIETIES

I have never seen any distinctive or named cultivars of false aralia. Note that its juvenile foliage is completely different from its mature foliage. The juvenile foliage is more delicate and lacy looking, with numerous speckles on leaves, petioles, and stems. The mature foliage—much larger, with a broader, lobed leaf—looks something like a brown version of *Fatsia japonica*.

PROPAGATION

False aralias may be propagated by cuttings and air-layering, though seed production is by far the most popular means for commercial producers. The fruit is collected from mature, flowering specimens in the landscape. The seeds are usually separated from the cleaned fruit, then planted in cell trays, frequently around 30 seeds per cell. Seeds are generally sown in a peat-lite type of mix with very good aeration and drainage. Some growers prefer to cover the top of the cell trays with either perlite or vermiculite. The seeds should be barely covered and kept evenly but not excessively moist until germination.

Avoid keeping young false aralia seedlings excessively wet, as they are rather susceptible to damping-off, specifically from *Pythium* and *Rhizoctonia*. Most growers apply a preventative drench at about the time of seeding, typically with Subdue (metalaxyl) and either Chipco 26019 (iprodione) or one of the thiophanate methyl fungicides. Seedlings should be fertilized very sparingly.

CULTURE

Unless you have a high-quality potting medium, keeping false aralia properly irrigated can be tricky. The plant does not appreciate extremes of wet or dry conditions. It prefers even moisture to avoid leaf drop from drought or root loss. Therefore, try to go with a somewhat lower moisture-holding capacity in the medium for false aralia. Avoid mixes with excessive amounts of peat, though certainly up to about 40% or so is fine. Definitely do not overwater this crop, or trouble will ensue.

Light levels are typically 2,000 to 4,000 f.c. (21.6 to 43.2 klux). Air movement is very desirable with this crop, so keep fans going. In dry areas keep the relative humidity above 25%, or problems with leaf tip injury can occur. This can be difficult in cold areas, where heating can dry out the greenhouse atmosphere. *Dizygotheca* tolerates temperatures below 45F (7.2C) without injury; most growers do not like to let them get below 40F (4.4C). The plant is somewhat heat tolerant, though leaf drop can occur at high temperatures. The most favorable temperature range is 70 to 80F (21 to 26.7C).

NUTRITION

False aralia fertility requirements are generally on the low side. These are somewhat slow-growing plants, so they cannot take advantage of large amounts of granular fertilizers at any given time, though for larger pot sizes, granular fertilizers are commonly used. It has been demonstrated that false aralias grow

TABLE 10 Leaf analysis rating standards for *Dizygotheca elegantissima*

Nutrient (%)	Very low	Low	Medium	High	Very high
Nitrogen	<2.2	2.2–2.4	2.5–3.5	3.6–4.2	>4.2
Sulfur	<0.18	0.18–0.24	0.25–0.50	0.51–0.75	>0.75
Phosphorus	<0.18	0.18–0.24	0.25–0.60	0.61–1.00	>1.00
Potassium	<1.40	1.40–1.75	1.80–3.50	3.55–4.25	>4.25
Magnesium	<0.25	0.25–0.29	0.30–0.40	0.41–0.75	>0.75
Calcium	<0.30	0.30–0.49	0.50–2.00	2.01–2.50	>2.50
Sodium			<0.11	0.11–0.20	>0.20
(ppm)					
Iron			<50–300	301–500	>501
Aluminum			<101	101–200	>200
Manganese	<40	40–49	50–300	301–500	>500
Boron	<20	20–24	25–50	51–75	>75
Copper	<4	4–5	6–200	201–400	>400
Zinc	<16	16–19	20–200	201–400	>400

Sources: Institute of Food and Agricultural Sciences, Apopka, Florida; Dr. Benjamin Wolf, Fort Lauderdale, Florida.

Notes: Common name: false aralia. Sample of most recent fully mature leaves, no petioles.

better with trace elements incorporated into the potting medium. N-P$_2$O$_5$-K$_2$O ratios of 2-1-1 or 3-1-2 are generally used. When incorporating slow-release, coated fertilizers, incorporation rates are generally in the low to medium range.

Dizygotheca plants don't have any particular nutritional requirements. They run low on magnesium at times as they mature, causing a yellowing toward the tips of the older leaves. Growers occasionally include magnesium sulfate, magnesium nitrate, or chelated magnesium in the spray program when growing it in larger pot sizes. Otherwise, nutritional deficiency symptoms are relatively rare.

A pesticide droplet may bounce up to six times before sticking to a leaf.

DISEASES

While numerous diseases of false aralia have been reported, there are really only three of consequence that growers are likely to encounter: *Alternaria, Rhizoctonia,* and *Pythium. Alternaria* causes leaf spots and leaf blight when the plant is grown under shadehouse conditions or when condensation from greenhouse roofs is prevalent. Foliage fairly quickly turns a bright canary yellow, then very dark brown. The leaves then drop. The symptoms look very much like drought stress or soluble salt injury. Sprays of Chipco 26019 (iprodione) or Dithane (mancozeb) are very helpful.

Rhizoctonia causes root rot as well as severe blight of older leaves. Roots generally become discolored, followed by severe browning and dropping of older foliage, especially under warm conditions. If you look carefully, you may see reddish brown threads connecting the fallen leaves to the top of the potting medium. *Pythium* may also be found when *Rhizoctonia* root rot is prevalent. In addition to drying the plants down, give soil drenches with Chipco 26019 (iprodione) or Terraclor (PCNB).

False aralias are also rather sensitive to nematode infestations, particularly those of root knot *(Meloidogyne)* and lesion *(Pratylenchus)* nematodes. Symptoms in false aralia include discolored roots, lack of root vigor, and stunted and chlorotic foliage. Plants produced from well-grown seedling liners should be free of the disease, but be careful not to place liner trays or plants on exposed ground, or infestation can result. Bench-grown plants generally do not have nematode problems, while those on the ground may be exposed to nematode populations. At this writing there are no chemical nematicides registered for ornamental plants in most of the United States.

Insect and mite pests

Scale and mealybug problems can be fairly common with dizygothecas. Liners should be inspected carefully when received, and the crop should be monitored periodically for the presence of these pests. Sprays of Dycarb (bendiocarb) are helpful against both insect pests. Horticultural oils can also be used in cool conditions, but be careful because false aralias are somewhat sensitive.

The major pest problem, however, is mites, both two-spotted mites (*Tetranychus)* and microscopic mites. As usual, hot, dry weather tends to favor mite infestation. Severe two-spotted mite infestation results in numerous white speckles on the foliage, and webbing may be observed. Microscopic mites generally cause distortion of the new foliage.

Predator mites may be effective in some situations in controlling two-spotted mites. Growers also tend to spray preventatively with Avid (abamectin), Pentac (dienochlor), or insecticidal soaps. Microscopic mites can be controlled with these miticides, as well as with Kelthane (dicofol). Use a wetting agent or surfactant when spraying for microscopic mites.

Disorders

Leaf drop with this plant can be related to moisture fluctuation, soluble salts, or ethylene. Dizygothecas are rather susceptible to elevated soluble salts, resulting in tip burn and leaf drop. If liquid feeding, make sure you schedule a leach periodically, and if using dry fertilizers, don't overapply. When plants are grown close together and air movement is poor, wet soil conditions can generate ethylene, causing leaf distortion and leaf drop. Unvented heaters can also generate ethylene.

Plants treated with Vydate (oxamyl) develop very narrow, strappy leaves, giving something of a witches'-broom effect. Vydate is no longer labeled on false aralia in the United States, and it should not be used on this plant under any circumstances. Finally, when moisture fluctuations are extreme, mature false aralias may develop edema symptoms, small, oval, reddish spots on the older foliage. It is a physiological disorder and really can be prevented only by managing your moisture carefully.

Tricks

The main tricks in producing false aralias successfully are to choose a good soil mix with a fairly low moisture-holding capacity, then manage your irrigation diligently. These steps alone will take care of 90% of your problems. If leaf drop does occur, the plants can be cut back to within 2 inches (5 cm) of the soil line and grown again. However, in shadehouse conditions or under high humidity, additional disease may continue to attack the stems, causing further dieback. If you have to cut stems back, spray fungicides afterward.

Chipco 26019 (iprodione) is a very good chemical for growing false aralias, and the grower should use it periodically as both a drench and a foliar application. Many growers like to drench with Chipco 26019 and Subdue (metalaxyl) at planting, frequently with a high-phosphate starter fertilizer to stimulate root growth. The plant does not have a very aggressive root system, so doing what you can to maintain root health is essential. Buyers of false aralias should check the root health before purchasing, if possible, as interior quality will be improved and extended if roots are healthy.

INTERIOR CARE

Humid conditions are preferred, so try to avoid relative humidity below 25%. In difficult situations it may help to place plants above a pan of moistened gravel in order to keep humidity up around them. Watch periodically for spider mite infestations.

As plants remain in interior environments, over time the foliage tends to become thin toward the bottom. This is partly due to water stress factors and partly due to the natural shedding of older foliage as the plants mature. Leaf drop is reduced under brighter interior conditions. False aralias prefer relatively high interior light, 320 to 1,000 f.c. (2.7 to 10.8 klux).

REFERENCES

Blessington, T.M., and P.C. Collins. 1993. *Foliage Plants: Prolonging Quality*, 92-93. Batavia, Ill.: Ball Publishing.

Chase, A.R. 1987. *Compendium of Ornamental Foliage Plant Diseases*, 74. St. Paul, Minn.: American Phytopathological Society.

Henderson, S., and C. Whitcomb. 1984. Response of False Aralia to Micronutrients and Osmocote. *Nurserymen's Digest* (February): 51.

McConnell, D.B., and R.W. Henley. 1983. False Aralia. *Foliage Digest* (August): 9.

Watkins, J.V., and T.J. Sheehan. 1975. *Florida Landscape Plants*, 303. Gainesville: The University Presses of Florida.

DRACAENA DEREMENSIS
Janet Craig and Warneckii

HABITAT

The genus name *Dracaena* comes from the Greek for "dragon." The first dracaena known to the Western world was *Dracaena draco*, whose red sap was associated with dragon's blood. Dracaenas are in the Agave family. There are about 40 species, mostly African, with a few species from Asia and one from Central America. Dracaenas are generally understory plants in tropical forests. Plants of this type tend to favor low light, high humidity, and warm conditions. *Dracaena deremensis* can reach a height of 15 feet (4.5 m) in its native Africa, but elsewhere it is usually much smaller.

USES

D. deremensis cultivars Janet Craig and Warneckii are excellent low-light interior plants. Their long, tapered leaves, pleated foliage, and rich colors help make them attractive, durable foliage plants. They are frequently used as floor plants in interior situations or for mass plantings in beds.

Much of the production is in 10-inch

Fig. 18 *Dracaena deremensis* cuttings that are rooting in Costa Rica. (Courtesy of Bill Lewis)

114

(25-cm) containers, with three tip cuttings per pot. Fourteen-inch (35-cm) pots are also produced, typically with three to four plants per pot. *D. deremensis* varieties are also produced in 6-inch (15-cm) containers, usually with one plant per pot. You occasionally see larger branched or character specimens, which are usually grown by rooting or transplanting parts of old stock plants.

VARIETIES

D. deremensis cv. Warneckii has stiff, tapered leaves with gray, green, and white stripes. Jumbo Warneckii is a very similar but larger and more robust version of Warneckii. Lemon Lime is an attractive Warneckii cultivar with more yellow and green in the stripes. Warneckii has a fragrant pink bloom, but I have seen it flower only once.

   D. deremensis Janet Craig is a sport of Warneckii with broad, shiny, deep green foliage. The leaves appear ribbed and are more flexible than Warneckii's. The plant was named for the daughter of nurseryman Robert Craig. Janet Craig Compacta is similar to Janet Craig but much smaller.

PROPAGATION

The majority of Janet Craig and Warneckii production is from tip cuttings, which are frequently produced in stock farms in the Caribbean, Central America, and elsewhere. Cuttings can be purchased in various sizes, usually unrooted. The cuttings are customarily planted directly in the finishing container. Rooting hormones are generally not needed, though they do speed rooting slightly. Mist is usually not used, though in hot weather one or two brief mistings around midday may be helpful. The cuttings need warm temperatures and moist soil. Rooting takes approximately three weeks, and only a small percentage of the cuttings fail to root.

   Air layers are also produced, frequently by cutting a notch on each side of a stem. Frequently, a pebble or similar small object is placed in the notch to keep it from healing over. Moist sphagnum moss is placed around the cut stem tissue, and the sphagnum is then wrapped in aluminum foil or sometimes clear polyethylene. Today the air-layering technique is more commonly used for larger, branched stems.

CULTURE

Most references suggest that *D. deremensis* cultivars should be grown between 2,000 and 4,500 f.c. (21.5 and 37.7 klux). In warm climates, however, you do better growing them between 1,500 to 1,800 f.c. (16.1 to 19.4 klux). You are shading Warneckii and Janet Craig more for temperature control than for actual light reduction. The higher the altitude your farm is at, the less shade they require.

D. deremensis cultivars do not like heat. The recommended maximum temperature is 90F (32.2C). As temperatures increase above 95F (35C), problems with chlorosis and leaf notching may develop. Many growers benefit from putting up an extra piece of 30 or 40% shade in the summer. The plants grow very little below 70F (21.1C). Ideal soil-root temperatures are between 75 to 80F (23.9 to 26.7C). Cold damage will occur around 35F (1.7C) or if plants are exposed to 55F (12.8C) for a week.

Janet Craig and Warneckii need a well-aerated potting medium, frequently a mixture of peat and pine bark with perhaps 10% sand. Their moisture requirements are about average, but it is best to avoid wet or dry extremes. The grower will do better growing this crop on the dry side. Most dracaenas are fluoride sensitive, and it is desirable to lime the medium to a pH of 6.5 and to maintain a good soil calcium level.

NUTRITION

Janet Craig and Warneckii cultivars usually have high requirements for phosphorus and iron, and they are sensitive to boron and fluoride. A common practice is to fertilize as a topdressing with a slow-release 18-6-8, 19-6-12, or similar ratio of fertilizer. Granular fertilizers high in iron can also be used to fertilize dracaenas, as can liquid fertilization of 200 ppm nitrogen from 9-3-6 about once a week. *D. deremensis* cultivars are somewhat sensitive to soluble salts and are fairly slow growers, so avoid excessive fertility.

It is best to use fertilizer and irrigation water low in fluorides. Experienced growers tend to use little or no boron in the fertilizer program for these plants. When fertility is too low, plants may develop weak color and rather narrow, strappy leaves. Lack of phosphorus can cause a severe dieback in middle and older foliage. Iron deficiency is quite common, resulting in a rather severe interveinal chlorosis, especially in Janet Craig.

DISEASES

Janet Craig and Warneckii have relatively few disease problems. Probably the most serious are root diseases, typically caused by *Fusarium* or *Pythium*. Janet Craig and Warneckii have what I call primary and secondary root systems. The primary roots are the thicker, trunklike roots. The secondary roots are the

TABLE 11 Leaf analysis rating standards for *Dracaena deremensis*

Nutrients (%)	Very low	Low	Medium	High	Very high
Nitrogen	<1.7	1.7–1.9	2.0–3.0	3.1–4.0	>4.0
Sulfur	<0.12	0.12–0.19	0.20–0.40	0.41–0.65	>0.65
Phosphorus	<0.12	0.12–0.19	0.20–0.40	0.41–0.75	>0.75
Potassium	<1.65	1.70–2.45	2.50–4.00	4.01–5.00	>5.00
Magnesium	<0.21	0.21–0.29	0.30–0.60	0.61–1.00	>1.00
Calcium	<0.80	0.80–0.99	1.00–2.00	2.01–3.00	>3.00
Sodium			<0.26	0.26–0.50	<0.50
(ppm)					
Iron	<30	30–49	50–300	301–500	>500
Aluminum			<151	151–500	>500
Manganese	<25	25–49	50–300	301–500	>500
Boron	<11	11–15	16–50	51–100	>100
Copper	<5	5–7	8–300	301–500	>500
Zinc	<15	15–19	20–200	201–400	>400

Sources: Institute of Food and Agricultural Sciences, Apopka, Florida; Dr. Benjamin Wolf, Fort Lauderdale, Florida.
Notes: Common name: dracaena Janet Craig. Sample of most recent fully mature leaves, no petioles.

smaller, fine-textured roots that originate from the primary roots. The grower should look after the health of these secondary roots, because if they were lost, plant quality would be greatly reduced. Many growers like to drench shortly after potting with a thiophanate methyl fungicide plus Subdue (metalaxyl) and a high-phosphate starter fertilizer.

The most common foliar disease is caused by *Fusarium moniliforne*, which was first observed in Hawaii. The leaf spots, generally dry and somewhat irregular in shape, primarily affect newer leaves. The edge of the leaf spot is generally reddish. Sprays of Daconil (chlorothalonil) or Dithane (mancozeb) help. Try also to irrigate when drying conditions are good.

Stem rots can be caused by *Erwinia* or *Fusarium*. The *Erwinia*, more of a wet stem rot, is frequently accompanied by a foul odor. Dips or drenches with Agrimycin 17 may be helpful in preventing *Erwinia*. *Fusarium* stem rot can be controlled to some extent with thiophanate methyl fungicide drenches. If excessive force is used in inserting a cutting into the medium, the cutting may rot.

INSECT AND MITE PESTS

D. deremensis has relatively few insect pests. Scales, mealybugs, and thrips are occasional problems. Mealybugs are identified by their white, cottony masses, which may move slowly. Several repeated sprays of Dycarb (bendiocarb), Orthene (acephate), Dursban (chlorpyrifos), or Enstar (S-kinoprene) should be helpful. Good coverage is critical.

Florida red scale may also attack *D. deremensis* cultivars. The scale infestation frequently but not always comes from the cutting source. The few initial scale insects will multiply as the crop matures, being generally found on the old leaves. Controls are the same as for mealybug, but granular Di-syston (disulfoton) is also very good. Thrips activity is rare, but it may appear as scarred leaves, especially the newer foliage. Thrips do not chew per se, but they have rasping mouth parts that scar foliage. Sprays with Mavrik (tau-Fluvalinate) or Orthene (acephate) should control them.

DISORDERS

Probably the most serious and troublesome disorder, especially in Janet Craig, is chlorosis, commonly known as netting. The new foliage becomes severely yellow, with very distinct transverse, green veins. It tends to occur later in the production cycle. Netting is worse in hot weather, but it can happen anytime. Keep the root system healthy; drenches with high rates of chelated iron may reduce the netting. In order for the iron to work, however, you need adequate soil aeration and a reasonable root system.

Notching is a rather curious disorder that affects Warneckii. The basal part of the leaf near the stem appears to have been slit several times by a razor blade. The disorder is worse under high light and high temperature conditions, though it can happen more or less anytime. Light foliar sprays of boron usually alleviate this symptom.

Both Janet Craig and Warneckii cultivars are sensitive to boron and fluoride toxicity. Fluoride creates several different symptoms in Warneckii. You may see elongated, brown leaf spots in the white-striped tissue. Orange blotches may develop, as may tip burn and marginal burn. Boron toxicity in Janet Craig tends to appear as a tip burn or marginal burn, usually near the leaf tip. Most foliage plants with long, tapered leaves tend to deposit undesirable substances, such as fluoride, in the tips of the old leaves. Avoid heat and moisture stress, keep soil pH and calcium levels up, and use low-fluoride fertilizers and irrigation water to the extent possible.

Boron toxicity symptoms are somewhat similar to fluoride injury symptoms, though they tend to be more concentrated near the leaf tips. Fluoride toxicity in Janet Craig appears as tip burn on older leaves, with orange flecks just in from the leaf tip. Many *D. deremensis* cultivars develop narrow, strappy leaves when root systems or soil structure are poor. In very wet weather Janet Craigs develop whitish and water-soaked spotting in the whorl. It usually clears up when dry conditions return.

TRICKS

Try to grow Janet Craig and Warneckii in a tall greenhouse for better air movement and heat dissipation. Shade the plants adequately for temperature control, and you may need to hang up additional shade or paint plastic roofs in summer. Janet Craigs, especially, have a hard time making chlorophyll at high leaf temperatures. If you gently grab the third leaf back from the growing tip during the middle of the day and hold it lightly in your hand, you may be able to feel the leaf's heat. If you can, this probably indicates that light and temperature are too high. Irrigate during the middle of the day to increase humidity and reduce leaf temperature. In hot conditions it may be helpful to apply mist briefly once or twice a day just to cool things off. Remember that you are shading as much for temperature control in *D. deremensis* cultivars as anything else.

> *Foliar-applied boron is distributed to all of the leaves, whereas root-absorbed boron may be dumped into older leaves.*

Use low-fluoride fertilizer and water sources, and if you have perlite in the medium, it should be leached prior to blending in the mix. The grower should inspect the secondary root system periodically and make sure the fine roots are in good health. If they begin to decline, it is a sign that you may need to drench with fungicides and reduce irrigation. Drenching with Subdue (metalaxyl) plus fairly high rates of EDDHA iron may help with root health and plant color.

If a newly planted cutting begins to rot from the base, recutting the stem into healthy tissue will usually not save the cutting. It should be discarded. The miticide Omite (propargite) may not be safe on *D. deremensis*. There is no need to spray miticides on these plants, anyway.

INTERIOR CARE

Try to purchase Janet Craig and Warneckii plants that have healthy root systems for good interior performance. *D. deremensis* cultivars do well under low light of 100 to 150 fc (1.5 to 1.6 klux). They tolerate 50 fc (0.5 klux), but a little more light is better. Exposure to low light conditions for extended periods causes subsequent emerging leaves to become more and more narrow. They don't need a lot of light, but keeping them fairly near a window will help, especially in northern climates. Avoid exposure to temperatures below 50F (10C) for any extended period.

Janet Craig and Warneckii tolerate moisture extremes pretty well, but if the plants get too dry, tip burn is likely to develop. If the plants are too wet, the leaves become very thin and strappy, and significant chlorosis or even cutting death may result. Try to maintain at least 40% relative humidity to avoid tip burn problems. It is best to let the plants dry a bit between irrigations.

If you are reasonably adept at watering, these dracaenas will tolerate a fairly wide variety of soils. Fertilize very sparingly with soluble triple 20 or 20-10-20, perhaps only every three to four months. Water quality is important. Try to avoid fluoridated tap water; well water or rainwater is generally preferable.

REFERENCES

Conover, C.A., and R.T. Poole. 1973. *Factors Influencing Notching and Necrosis of* Dracaena deremensis *'Warneckii' Foliage.* Fla. Agric. Exp. Sta. Journal Series 4941: 378-384. Apopka: University of Florida Agricultural Research Center.

McConnell, D.B., and R.W. Henley. 1983. Warneck Dracaena. *Foliage Digest* (October): 11.

Poole, R.T., A.R. Chase, and L.S. Osborne. 1985. Dracaena *'Warneckii'* and *'Janet Craig'.* AREC-A Foliage Plant Research Note RH-1985-C. Apopka: Agricultural Research and Education Center.

Poole, R.T., and C.A. Conover. 1988. Heat Tolerance of Dracaena 'Janet Craig' and Dracaena 'Warneckii'. *Nursery Digest* (October): 24.

Watkins, J.V., and T.J. Sheehan. 1975. *Florida Landscape Plants*, 103. Gainesville: The University Presses of Florida.

DRACAENA FRAGRANS
Corn Plant

HABITAT

Dracaena fragrans has been in use as an indoor plant in Europe since the mid-1700s and in the United States since the early 20th century. The name *fragrans* refers to the very sweet-smelling flowers which are occasionally produced.

Commonly called the corn plant, *Dracaena fragrans* is native to the African region of Upper Guinea, where it grows in humid, tropical forests. In its native habitat the plant is frequently but not always shaded by surrounding vegetation. The plant's medium-green leaves average 3 feet (0.9 m) in length and are generally about 4 inches (9 cm) wide.

Fig. 19 *Dracaena fragrans* on the left, *D. fragrans* cv. Lindenii on the right. (Photo by the author)

USES

Corn plants are produced by commercial growers in a variety of configurations. Single tip cuttings are grown in 6-inch (15-cm) containers, and three cuttings in a 10-inch (25-cm) pot make an attractive, bushy floor plant. *D. fragrans* is also marketed as multiple sprouted canes, typically with various sizes planted in the same pot. The resulting staggered heights mean foliage sprouts up and down the cane grouping in an even fashion. This makes *D. fragrans* excellent as a tall indoor specimen.

Sometimes stock plants are harvested and planted in containers, where they send out multiple heads, for an unusual effect. Branched cane pieces are also grown in stock farms in Central America and the Caribbean. These branched canes are then assembled to make attractive large-container plants. The tops of the cane pieces are generally dipped in paraffin or a mixture of wax and concrete in order to retard moisture loss. The plants are also used as tall specimens in landscapes, and single-cane pieces are planted to make living fences in many tropical countries.

VARIETIES

The true *D. fragrans* is olive green and nonvariegated. More popular is *D. fragrans* cv. Massangeana, which is generally referred to as the corn plant. Massangeana has broad, single, yellow stripes in the leaf. Lindenii has a similar form but a mostly yellowish leaf with variegated green striping near the midrib. The rare Victoria has abundant white and cream stripes.

PROPAGATION

Tip cuttings approximately 12 inches (30 cm) in length are generally planted directly in finishing containers. These may be placed in a tent to maintain humidity, but if in the open, the cuttings may require misting two or three times per day. Cane pieces may be rooted upright in sawdust or sand, but they are more commonly rooted directly in the growing pot. Cane pieces require little mist; too much, and trunk rot may ensue. Rooted and sprouted cane pieces of various lengths are sometimes sold by suppliers.

While not a common practice, it helps to cut off the lower one-half inch (1.25 cm) of the cane piece, which may then be dipped in a thiophanate methyl fungicide mixed with Agrimycin 17 (streptomycin sulfate). Indolebutyric acid (IBA) at approximately 5,000 ppm, applied to the base of the cane, may help with rooting. For increased rooting some growers like to cut the basal portions of the canes with a circular saw, making a one-eighth-inch (3-mm) cut every 2 ¹/₂ inches (6.25 cm). Extra roots will sprout from the wounded tissue, especially if IBA is added.

Cane pieces should have adequate maturity because green cane may break down and rot. Thicker cane pieces tend to send out more heads, whereas thinner cane pieces root better. Avoid planting old cane pieces. You can tell if cane is old, as it usually has an elliptical hollow spot on either end, resulting from drying of the woody tissue over time.

Most growers have to plant cane pieces fairly deeply and compact the soil to some degree to keep the canes vertical in the pot. Roots and sprouts generally form within three to six weeks of planting. Air layers of smaller cane pieces are also sometimes used.

CULTURE

The best light levels are generally 3,000 to 3,500 f.c. (32.4 to 38.7 klux). This is about 73% shade, though you can grow these plants in 63% in cooler situations. The actual light level required would depend on temperature, with more shade being required in hotter areas. As with *Dracaena deremensis,* the higher the altitude, the less shade required. When growing cane it is important to space the plants so that the small canes get adequate light. For this reason, it helps to have smaller cane pieces oriented toward an aisle so they will not be shaded by larger cane pieces.

Optimum temperatures are 65 to 96F (18 to 35C). Try not to expose the plants to temperatures below 55F (13C), or growth and quality may be reduced. Cold injury occurs as temperatures approach 40F (4.4C), especially if it is windy. High temperatures contribute to leaf burn.

The medium's pH range is generally 6 to 6.5, though keep it on the high side when fluorides are a problem. The medium should have a little higher bulk density than normal in order to keep plants upright. Frequently, 10 to 15% sand is incorporated into mixes made mainly of peat and either bark or other wood sources. If perlite is used, it should be leached to remove fluorides prior to incorporation in the mix.

Protect the plants from wind and cold. Fan-and-pad cooling is helpful in hot greenhouses. Ideally, *D. fragrans* should be grown in a tall, cool, well-ventilated greenhouse. If humidity is low, it helps to leave buckets of water out, or else periodically hose down the ground to maintain humidity.

NUTRITION

The typical foliage ratio of 3-1-2 $N-P_2O_5-K_2O$ is generally used with *D. fragrans,* though fertilizer sources should be low in fluoride. Superphosphate sources are generally discouraged because of potential fluoride content.

Fertilization should be gentle, as the corn plant can burn rather easily and is sensitive to soluble salts and ammonia. Most growers use minimal boron in their fertilizers, as well. It is generally best to wait until cane pieces root before applying fertilizer. Cane plants frequently do not require fertilizer until later in the production cycle, as the cane pieces have plenty of stored nutrients within them.

If soil aeration is poor or root disease is present, iron deficiency may appear as bleaching in the upper foliage. Manganese deficiency is very similar to iron deficiency in appearance, though it is probably more common. When a leaf is lacking manganese, the transverse veins may still appear green, while the rest of the leaf is chlorotic. Lack of phosphorus shows as severe leaf dieback from the tip, frequently making one-third to one-half of the leaf necrotic. Necrotic leaf tissue also tends to leak potassium severely. When cane is low in boron, small, stubby, distorted, multiple heads tend to form. Foliar boron sprays fix the problem.

TABLE 12 **Leaf analysis rating standards for *Dracaena fragrans* cv. Massangeana**

Nutrient (%)	Very low	Low	Medium	High	Very high
Nitrogen	<1.9	1.9–2.1	2.2–3.5	3.6–4.5	>4.5
Sulfur	<0.12	0.12–0.19	0.20–0.75	0.76–1.25	>1.25
Phosphorus	<0.11	0.11–0.14	0.15–0.40	0.41–0.75	>0.75
Potassium	<1.500	1.50–1.95	2.00–4.00	4.05–5.00	>5.05
Magnesium	<0.15	0.15–0.20	0.21–0.75	0.76–1.00	>1.00
Calcium	<0.80	0.80–0.99	1.00–2.50	2.50–3.50	>3.50
Sodium			<0.26	0.26–0.50	>0.50
(ppm)					
Iron	<30	30–49	50–300	301–500	>500
Aluminum			<151	151–500	>500
Manganese	<30	30–49	50–300	301–500	>500
Boron	<11	11–19	20–50	51–100	>100
Copper	<5	5–7	8–50	51–200	>200
Zinc	<15	15–19	20–200	201–400	>400

Sources: Institute of Food and Agricultural Sciences, Apopka, Florida; Dr. Benjamin Wolf, Fort Lauderdale, Florida.

Notes: Common name: mass cane. Sample of most recent fully mature leaves, no petioles.

Diseases

When cane pieces rot, the causal organism is usually either a *Fusarium* or an *Erwinia* species. *Fusarium* typically appears as a dry, dark rot, frequently with loosened bark. *Fusarium* also occasionally causes sprouted heads to turn charcoal colored. Cane infected with *Erwinia* is generally wet and mushy, frequently with a foul odor. Dips or drenches with a thiophanate methyl fungicide and Agrimycin 17 (streptomycin sulfate) may help prevent these diseases.

Fusarium may also cause a large, dry, tan leaf spot. Keeping foliage dry in conjunction with sprays of Daconil (chlorothalonil) will help. In hot, wet situations, *Rhizoctonia* blight may develop on plants grown from tip cuttings. Numerous charcoal-colored blotches form very suddenly on the foliage. Cane plants are generally not affected, but if *Rhizoctonia* occurs, sprays of Daconil (chlorothalonil) or a thiophanate methyl fungicide should stop the infection.

Insect and mite pests

Two species of ambrosia beetles *(Xyleborus)* attack *D. fragrans* cane, especially during the rooting process. The insect looks like a tiny cockroach about one-sixteenth of an inch (1.6 mm) long. It makes numerous holes in the cane, about the size of holes you would make with a thumbtack. Sawdust or frass from the beetle activity is frequently seen at the base of the cane. Mature wood is generally attacked, and the presence of ambrosia beetle is usually indicative of cane that is weakened by other problems. Ambrosia beetle is found in most tropical areas of the world, and an ambrosia beetle infestation may or may not result from the cane source. Cane drenches with Lindane will help, but badly affected plants should be discarded.

The banana moth *(Opogona)* can also be very troublesome. The rarely seen moth generates large, segmented larvae, which feed on cane just under the bark. The black-headed larvae wiggle constantly. Sprays of Sevin (carbaryl) or the biological insecticide Biovector *(Steinernema carpocapsae)* are useful in controlling *Opogona*. Growers of *D. fragrans* will also occasionally encounter mites, scales, caterpillars, and grasshoppers. These infestations are relatively rare on this plant, however.

Disorders

Leaf tip burn is a major headache for corn plant growers. It can be caused by an accumulation of boron or fluoride, high soluble salts, or moisture stress.

Fluoride accumulation results in brown- or gray-colored leaf tips and margins. The necrosis progresses, frequently with a yellow halo. This problem is common in interior situations. In commercial nurseries the most susceptible stage for tip burn is when the plants are 60 to 75% grown. At this stage the shoots may be somewhat more developed than the root systems, although still immature. Accumulation of boron results in tip burn that is more tan in color, with less of a halo.

Iron chlorosis can develop when soil is low in oxygen or high in pH. Drenches or topdressing with chelated iron products may relieve chlorosis. If pale, weak heads form, it may indicate that the cane piece has failed to root. Narrow, strappy leaves can develop under low light conditions or when roots are weak. Massangeana tends to lose the variegated striping in its older foliage when nitrogen is low.

In hot weather *D. fragrans* twists its leaves sideways to reduce leaf exposure to the sun. Plants return to normal orientation when the heat subsides. Chilling injury can occur somewhere between 35 and 50F (1.7 and 10C), depending on the amount of wind that accompanies the cold. Cold injury appears as light gray or brown leaf margins, possibly with bands of necrotic tissue in the whorl. Consistently cool nights stimulate flowering of *D. fragrans*, which is bad news for growers. The flowers are attractive and fragrant but very short lived, and the growing point that has flowered will stop growing.

 Fluoride accumulation prevents proper closing of the stomates, causing tip burn.

TRICKS

To reduce tip burn, keep the nitrogen-to-calcium ratio low with calcium sprays. Regular sprays of calcium nitrate or chelated calcium are very beneficial. Occasional sprays with antitranspirants also help reduce leaf burn in hot or windy conditions.

One-foot (30-cm) cane pieces tend to accumulate at stock farms that are cutting larger cane pieces to size. The 1-foot canes may therefore be old or immature. If you are having trouble with dieback with 1-foot canes, it helps to buy 2-foot (60-cm) cane pieces and cut them in half yourself.

Inspect young canes regularly for the number of heads that are forming. Single-headed pieces can be made into multiples by breaking off the bud when it is about 1 inch (2.5 cm) long. The piece will generally send out several new heads afterward. If more than four heads start on a cane piece, thin them to three or four of the biggest buds. Too many heads would result in poor development and a greater incidence of tip burn.

When planting tip cuttings, hold the stem about 2 inches (5 cm) from the base of the cutting. Don't push on the cutting from the upper part, or the injury may result in cutting death.

Stock farm growers can strip cane of older foliage about a month before harvesting in order to speed cane maturation. Stock plants generally require very little fertilizer, though low pH should be avoided. Be careful of temperature fluctuations around shipping time. Large temperature differences between the greenhouse, the truck, and the storage facility have created numerous headaches for interiorscapers.

Finally, cutting notches into the sides of cane pieces helps induce breaks at desired locations. With a hacksaw, cut about one-third of the way into the cane above a node, and you will generally get a head formed right below the notch.

INTERIOR CARE

It helps to maintain corn plants indoors if you can keep them near a window or with curtain-filtered sun. Corn plants can tolerate down to 50 f.c. (0.5 klux) for a time, though they may lose their striping and develop poor leaf expansion; 75 to 150 f.c. (0.8 to 1.6 klux) are preferred. Old corn plants in interiors generally develop long, stretched, thin leaves with very little variegation. You can cut corn plants back to any desired height if they become unattractive, and new buds will break. The tip cuttings can generally be rooted easily.

Corn plants may also require dusting from time to time, as dust may accumulate on the upper leaf surfaces and in the whorl. Try to use low-fluoride irrigation water; rainwater is excellent. Corn plants do best if relative humidity is 40% or more.

The plant will survive relatively poor interior conditions, though it may lose attractiveness with time. Maintaining decent light levels and occasional gentle fertilization will help extend useful interior life.

REFERENCES

(see references in preceding variety, *Dracaena deremensis*)

DRACAENA MARGINATA
Dragon Tree

HABITAT

Like all dracaenas, *Dracaena marginata* is a member of the Agave family. Common names include "red-edge dracaena" and "dragon tree," but in the trade the plant is generally referred to by its Latin name. *D. marginata* is native to Madagascar, off the eastern coast of southern Africa. The plant has woody stems, with narrow, sword-like leaves. The dark green, glossy foliage has narrow, red stripes near each leaf margin, hence the species name. *D. marginata* is fairly drought tolerant, thanks to the distinct wet and dry seasons in its native habitat. It also has an aggressive root system for the same reason.

Fig. 20 Ten-inch containers of *Dracaena marginata* (*left*) and *Dracaena marginata* Tricolor (*right*). (Photo by the author)

USES

Dracaena marginata is a versatile foliage plant with quite a number of potential uses. Selection may range from small tip cuttings in 3-inch (7.5-cm) pots to huge plants in large tubs. A common method of production for chain stores and

128

the retail market is to place three plants per pot in either 8-inch (20-cm) or 10-inch (25-cm) pots. This yields a fairly inexpensive, easy-to-grow houseplant.

D. marginata sends out numerous new shoots when cut back; therefore, branched character plants are frequently grown. These may come from field-grown stock plants or, in tropical areas, from mature landscape specimens. The European market likes smaller containers of character *D. marginata*, whereas in the U. S. market it is common to see character *D. marginata* in 14-inch (35-cm), 17-inch (42.5-cm), or even larger containers. Small tip cuttings are common in dish gardens and combinations. Plants can grow as high as 12 feet (3.6 m) in landscapes in USDA Zones 10B to 11.

VARIETIES

Like the standard variety, Bicolor has red and green stripes, but the variegation pattern is different. Colorama is similar to Bicolor, except there are red and ivory bands in the central part of the leaf. Tricolor has longitudinal, ivory stripes through the central green area and near the margins. Magenta is similar to the basic *D. marginata*, but more purplish to dark red overall. The more highly variegated types are only slightly less hardy and vigorous than the standard variety. Incidentally, *Dracaena reflexa* and its variegated sport Song of Jamaica are grown almost identically to *D. marginata*.

 Dracaena marginata is called the money tree in Hawaii because the first ones planted there were placed around the Bank of Hawaii.

PROPAGATION

Tip cuttings are generally planted directly in the container for smaller pot sizes. Some branched pieces may be air-layered with sphagnum moss and aluminum foil. Large character plants are typically planted directly into large containers; frequently they already have some roots. Mature *D. marginata* stock plants are often seen in the Caribbean basin countries as well as in Central America. Nurseries in tropical areas may maintain their own stock fields.

Indolebutyric acid helps with rooting, but it is generally not considered essential, as cuttings typically root in about three weeks anyway. Large cuttings may benefit from dips or sprays in antitranspirants prior to planting. Cuttings usually lose a few leaves during the rooting process.

CULTURE

D. marginata is fairly tolerant of soil types, but it is best to have a mix that is well drained yet moisture retentive. Various combinations of peat, bark, sawdust, and sand are used, depending upon local availability of medium ingredients. *D. marginata* tolerates low pH better than many other dracaenas, and an initial pH of 5.5 to 6 works well. The pH is best if a little bit low, as *D. marginata* has relatively high trace element requirements. It should be grown in fairly bright light, typically between 3,000 and 6,000 f.c. (32.4 and 64.8 klux). If grown darker, foliage may droop.

Optimum growing temperatures are between 75 and 90F (24 and 32.2C). It is preferable to keep night temperatures at 60F (16C) or above, though they can occasionally go somewhat cooler without injury. Plants tend to burn somewhere between 30 and 35F (-1.1 and 1.67C). *D. marginata* will be severely damaged or killed if temperatures drop below freezing for any period of time. It also helps to protect it from low-humidity situations and high winds.

NUTRITION

Red-edge dracaenas are usually fertilized with either a 2-1-2 or a 3-1-2 ratio of $N-P_2O_5-K_2O$. They are moderate to heavy feeders. All of the trace elements are extra important with *D. marginata,* so it is helpful to have trace elements incorporated into both the soil mix and the spray program. Granular fertilizers are commonly used as a topdressing, and smaller pots are frequently fertilized with soluble fertilizers at 200 ppm nitrogen in a constant liquid feed, with an occasional leaching. Slow-release fertilizers may be top-dressed or incorporated, especially for larger containers.

Lack of nitrogen results in pale, poorly expanded leaves. When sulfur is lacking, the plants grow normally but have rather weak color. Lack of phosphorus appears as a purpling of the older foliage. Potassium deficiency shows up as chlorosis, speckling, and necrosis in the older foliage. Plants lacking magnesium have older leaves whose margins turn canary yellow. Plants low in calcium are weak and may develop hollow stems if the deficiency is severe.

Iron deficiency in *D. marginata* shows as acutely chlorotic top leaves. Manganese deficiency is common, indicated by a less severe yellowing than with iron deficiency, with the veins remaining green. Plants lacking zinc have small, stunted leaves, one-half or less of their normal size. They may be somewhat withered looking and have dull color. Copper deficiency can be indicated by a tight, hard spear at the growing tip instead of the normal emerging shoot. The hard spear may break off. Water stress can cause a similar symptom. *D. marginata* lacking boron have leaves that droop gently, even in good light. Leaves may also be more susceptible to flecking.

TABLE 13   Leaf analysis rating standards for *Dracaena marginata*

Nutrient (%)	Very low	Low	Medium	High	Very high
Nitrogen	<1.9	1.9–2.	2.2–3.2	3.3–4.0	>4.0
Sulfur	<0.12	0.12–0.19	0.20–0.40	0.41–0.65	>0.65
Phosphorus	<0.15	0.15–0.17	0.18–0.40	0.41–0.75	>0.75
Potassium	<1.25	1.30–1.75	1.80–3.50	3.55–4.50	>4.55
Magnesium	<0.15	0.15–0.19	0.20–0.50	0.51–0.80	>0.80
Calcium	<0.60	0.60–0.99	1.00–2.50	2.51–3.50	>3.50
Sodium			<0.26	0.26–0.50	>0.50
(ppm)					
Iron	<30	30–39	40–300	301–500	>500
Aluminum			<250	251–2000	>2000
Manganese	<35	35–39	40–300	301–500	>500
Boron	<15	15–19	20–50	51–100	>100
Copper	<5	5–7	8–50	51–200	>200
Zinc	<15	15–19	20–200	201–400	>400

Sources: Institute of Food and Agricultural Sciences, Apopka, Florida; Dr. Benjamin Wolf, Fort Lauderdale, Florida.

Notes: Common names include marginata, cordyline. Sample of most recent fully mature leaves, no petioles.

DISEASES

The most common foliar disease of *D. marginata* is *Fusarium moniliforme*. The leaf spots are tan to reddish brown, usually with an orange or yellow halo. Sprays of Daconil (chlorothalonil) or Dithane (mancozeb) control it, and it also helps to maintain high fertility levels. *Cercospora* causes another fungal leaf spot. The tiny, red or dark green spots on the lower leaf surface are much smaller than those of *Fusarium*. Chemical controls are similar, though thiophanate methyl fungicides may also be effective.

Like *Dracaena fragrans*, *D. marginata* is affected by two primary stem rot diseases. *Fusarium oxysporum* causes dry, loose bark and darkened internal tissue.

This fungus may also attack stems after plants are cut back. Sprays of thiophanate methyl fungicides or Daconil (chlorothalonil) may be helpful. Bacterial stem and leaf rot caused by *Erwinia* may occur, especially during propagation or when rainfall or misting are excessive. The bacterium causes a wet, smelly, mushy rot. Cultural controls are best, including purchasing clean cuttings and avoiding excess moisture. Copper fungicides, Phyton 27 (picro cupric ammonium formate), and Agrimycin 17 (streptomycin sulfate) are also helpful. Lesion nematode may hollow out roots, especially in mature, field-grown plants. There are no current registered controls in most of the United States other than organic nematicides, such as Clandosan, Di-Tera, or Neotrol.

INSECT AND MITE PESTS

Florida red scale is fairly common in this dracaena. It tends to come in on cuttings from stock plants and will worsen with time as populations increase. Applications of granular Di-syston (disulfoton) provide excellent control, as do sprays of Dycarb (bendiocarb), Cygon (dimethoate), and others. Two-spotted mites cause speckling of foliage and eventually dull color. Sprays of Kelthane (dicofol), Mavrik (tau-Fluvalinate), Pentac (dienochlor), or Avid (abamectin) should control them.

Thrips may attack the growing tips, causing scarred or callused leaves. Sprays with Mavrik (tau-Fluvalinate) or Orthene (acephate) are helpful, especially when used with a wetting agent. Fungus gnats may feed on the bases of cuttings, contributing to disease and stem rot. Effective controls include Gnatrol *(Bacillus thuringiensis)* or Diazinon.

DISORDERS

Flecking is a common, serious disorder of *D. marginata*. Numerous white spots appear near the leaf tips of the emerging shoots. It is not caused by a pathogen, though *Fusarium* may attack the flecked spot as a secondary invader. The cause of flecking is said to be unknown, though research indicates that less flecking ensues if plants are grown in 80% shade and if light is reduced to less than 2,000 f.c. (21.5 klux). In my experience, though, flecking still happens at those light levels. Foliar sprays of boron often reduce flecking, and spraying other trace elements may help, as well. I suspect flecking may be a combination of nutritional and environmental factors, with carbohydrate metabolism probably involved.

Chlorotic bands across several leaves in the whorl may occur when plants are exposed to chilling temperatures. Cold injury also appears as rusty, jagged blotches on the edges of older leaves. Leaf tip burn may be caused by accumulations of excessive fluoride or boron or from high soluble salts. When the emerging growing tip or spear is tight and hard, drought stress or copper deficiency is usually the cause. Numerous small, yellow spots may form on the exposed part of the leaf base where it joins the stem. I suspect this may be some sort of guttation, as a sticky substance is frequently exuded from the spot. Ants and flying insects are attracted to this substance, which can make for a problem for stock plant producers.

TRICKS

In order to reduce winter flecking and increase winter plant quality, it helps to spray with trace element preparations during October or November. Boron sprays may be helpful against the flecking symptom in some cases, but not all. Foliar sprays of copper frequently help loosen tight spears. Boron and fluoride toxicity are less of a production problem than with other dracaenas and cordylines, probably because the symptoms are less severe. Sprays of Ornalin (vinclozolin) are reported to cause chlorosis.

The Tricolor and Bicolor varieties are less susceptible to *Fusarium* leaf spot. Large plants grown from tip cuttings may be hard to wet when irrigating overhead, as the old leaves deflect the irrigation. In these cases it helps to get the soil wet by either hand-watering, drip irrigation, or ebb-and-flow irrigation. Drought-stressed plants tend to form tuberlike growths on the roots. Use extra manganese in the fertilizer program for best leaf color.

INTERIOR CARE

The best room temperatures for *D. marginata* range between 65 and 80F (18 and 27C). At least 200 f.c. (2.2 klux) of light is best, though the plants will tolerate 75 to 100 f.c. (0.8 to 1.1 klux). However, the plants may lose older leaves and color in time if light levels are low.

Allow them to dry slightly between waterings, then irrigate thoroughly. Do not allow the saucer under the pot to accumulate water, or root rot will likely occur. Growth indoors is usually very slow, so repotting is rarely needed. The aggressive roots may eventually, break the container, however.

Be careful of fluoride levels in the irrigation water. Include trace elements in the fertilizer program from time to time. *D. marginata* can be fertilized lightly in interiors if light levels are good.

REFERENCES

Blessington T.M., and P.C. Collins. 1993. *Foliage Plants: Prolonging Quality,* 107-110. Batavia, Ill.: Ball Publishing.

Chase, A.R. 1987. *Compendium of Ornamental Foliage Plant Diseases,* 82.St. Paul, Minn.: American Phytopathological Society.

Chase, A.R. 1990. Fertilizer Rate Affects Severity of Fusarium Leaf Spot of Red-Edge Dracaena. *Foliage Digest* (June): 6-7.

Chase, A.R. 1993. Common Diseases and Disorders of Dracaenas. *Landscape and Nursery Digest* (October): 33-34, 74.

Osborne, L.S, A.R. Chase, and R.W. Henley. 1984. Dracaena marginata. AREC-A Foliage Plant Research Note RH-1984-D: 1-7. Apopka: Agricultural Research and Education Center.

Poole, R.T., and C.A. Conover. *Flecking of* Dracaena marginata. AREC-A Research Report RH 85-8: 1-2. Apopka: Agricultural Research and Education Center.

DRAGON TREE

(See Dracaena marginata*)*

DWARF SCHEFFLERA

(See Schefflera arboricola*)*

EMERALD GEM

(See Other Foliage: Homalomena*)*

ENGLISH IVY

(See Hedera*)*

EPIPREMNUM

Pothos

HABITAT

One of the most familiar foliage plants, golden pothos is native to the Solomon Islands in the Pacific. It was originally named *Pothos aureus,* but in the mid-1900s the genus was changed to *Scindapsus,* then finally to *Epipremnum.* Pothos is still the most widely used common name, although it is also called devil's ivy, ivy arum, and several other names.

Pothos is a true foliage plant, as it rarely flowers in culture. In fact, I have never seen a pothos flower. The juvenile leaves are heart shaped, with marbled patterns of green, yellow, and white. The average leaf is about 3 inches (7.5 cm) across, though leaves can become as large as 12 inches (30 cm) in diameter. The leaves tend to become much larger when the plant is growing upward on a tree or stake. In fact, in nature pothos leaves can become as large as 30 inches (75 cm) across. The mature foliage frequently has deep perforations between the veins, similar to *Monstera.* In its native habitat of humid, tropical forests, pothos grows both along the ground and up tree trunks.

USES

Pothos is one of the most durable and tolerant foliage plants. Numerous cuttings are planted per pot to

Fig. 21 Marble Queen pothos production, Delray Beach, Florida. (Courtesy of Ed Clay)

135

achieve fullness. Small containers between 3 and 6 inches (7.5 and 15 cm) are produced in great numbers. Hanging baskets are very popular, generally in sizes ranging from 6 to 10 inches (15 to 25 cm). Medium and large totems are also produced, where cuttings are rooted in the soil and are trained to grow up a vertical support made of wood, styrofoam, or other material.

VARIETIES

The most common *Epipremnum* is the golden pothos, with interwoven layers of green, yellow, and white. Hawaiian is a selection of golden pothos with rich variegation and somewhat more yellow pigmentation than the standard. Marble Queen has green and white, and the petioles are also variegated. There are at least three solid green pothos varieties, including Green Gold, Jade, and Tropic Green.

PROPAGATION

Almost all pothos propagation is from leaf-and-eye cuttings. The cutting consists of a leaf with a single stem node, a small piece of root emerging from the node. Stock plants are frequently grown in ground beds or raised beds. Growers with more limited space tend to trim their hanging baskets and use them as cutting sources, rather than maintaining stock beds. Some particularly astute growers maintain stock beds of pothos with different characteristics (leaf size, stem caliper, color, and so on), depending on the size of plant to be grown. High light levels (5,000 f.c., 57 klux) generally help with stock plant quality.

The best eyes are toward the young end of the vine. The stem of the cutting should be from 1 to 1.5 inches long (2.5 to 4 cm). It is best to leave at least one-fourth inch of the node root intact.

Cuttings are generally misted for one to two weeks, but not longer. The amount of mist depends on the greenhouse environment. Try not to lose the eye leaf, or the cutting will grow slowly. Avoiding moisture stress helps the cutting retain the eye leaf and the energy it provides.

CULTURE

The best light levels for pothos production range between 1,500 and 3,500 f.c. (16.1 and 37.7 klux). Light levels below 1,000 f.c. (10.8 klux) result in loss of variegation. Preferred growing temperatures are 70 to 90F (21.1 to 32.2C).

Pothos tolerates temperatures up to 95F (35C), but quality is reduced once temperatures go above 100F (37.8C). Maintaining night temperatures of 70F (21.1C) provides much better growth and quality than 60F (15.6C) nights. Foliage may start to yellow as temperatures drop below 55F (12.8C). Marble Queen is more cold sensitive than golden pothos, especially when it has a high degree of variegation.

Potting media containing sphagnum peat and bark are popular, sometimes with added perlite or polystyrene beads. Some pothos growers do well with straight sphagnum moss as a potting medium. The ideal soil temperature is 82F (27.8C). Plants grow very slowly in cold soil. Production time has been shown to be 35% shorter if bottom heat maintains a soil temperature of 70F (21.1C) in winter. Bottom heating is not necessary in summer. Irrigation frequency depends on medium and temperature, but mature pothos, especially, tends to require frequent irrigation.

NUTRITION

Pothos has fairly high fertility requirements, and magnesium should be emphasized. A minimum of 4 grams of a 3-1-2 slow-release fertilizer, such as 18-6-12, per 6-inch (15-cm) pot every three months has been successful; so has liquid feed of 200 ppm from 24-8-16, one to two times per week. Fertilizer rate in winter should be about 30% less than summer rates. One study showed that pothos requires about 100 mg of nitrogen per 6-inch pot per week. High-nitrogen fertility rates in stock production help cuttings grow faster after rooting. Nitrogen source does not appear to be important in pothos.

Extra potassium generally does not result in improved appearance, but you may get better growth and rooting of cuttings. Extra magnesium tends to help with variegation, but be very conservative on manganese fertilization. Low calcium results in thin, soft leaves, in which case calcium sprays may be very beneficial. Lack of potassium is not common, but when it does occur, internodes are short, and leaves become small, ultimately turning necrotic and dying.

DISEASES

Pythium splendens is one of the most common root rot diseases in pothos. When plants are wet and drainage is poor, the roots turn mushy and black. The stem may also turn black, and older leaves become solid yellow. The disease spreads

TABLE 14 Leaf analysis rating standards for *Epipremnum aureum*

Nutrient (%)	Very low	Low	Medium	High	Very high
Nitrogen	<2.25	2.25–2.69	2.70–3.50	3.51–4.50	>4.50
Sulfur	<0.15	0.15–0.19	0.20–0.50	0.51–1.00	>1.00
Phosphorus	<0.15	0.15–0.19	0.20–0.50	0.51–1.00	>1.00
Potassium	<2.40	2.40–2.99	3.00–4.50	4.51–6.00	>6.00
Magnesium	<0.25	0.25–0.29	0.30–0.60	0.61–1.00	>1.00
Calcium	<0.70	0.70–0.99	1.00–1.50	1.51–2.50	>2.50
Sodium			<0.20	0.21–1.00	>1.00
(ppm)					
Iron	<40	40–49	50–300	301–1000	>1000
Aluminum			<251	251–2000	>2000
Manganese	<30	30–49	50–300	301–1000	>1000
Boron	<15	15–19	20–50	51–75	>75
Copper	<4	4–5	6–50	51–300	>300
Zinc	<15	15–19	20–200	201–1000	1000

Sources: Institute of Food and Agricultural Sciences, Apopka, Florida; Dr. Benjamin Wolf, Fort Lauderdale, Florida.

Notes: Common names include pothos, *Scindapsus aureus*. Sample of most recent fully mature leaves, no petioles.

with splashing water or handling infected plants. Try to dry infected plants down, remove affected plant material, and drench with Subdue (metalaxyl), Aliette (fosetyl-aluminum), or a similar water mold fungicide.

Rhizoctonia root and stem rot was first described in pothos in 1955. Its appearance is similar to that of *Pythium,* though it may also cause leaf spots. Temperatures above 86F (30C) will reduce *Rhizoctonia* activity. Culturally, treat as for *Pythium,* and drenches of either Terraclor (PCNB) or Terraguard (Triflumizole) will be helpful. Do not drench with Chipco 26019 (iprodione).

There are two primary bacterial diseases common in pothos, *Erwinia* and *Pseudomonas. Pseudomonas* causes water-soaked leaf spots which spread quickly and lack any visible fruiting bodies. Do your best to keep the foliage dry, and spray either with a mixture of Dithane (mancozeb) plus a copper fungicide, or with Phyton 27 (picro cupric ammonium formate). Agrimycin 17 (streptomycin sulfate) is also commonly used, but it is not labeled for pothos in the United States. *Erwinia* causes a mushy, soft, smelly stem rot. This disease

FOLIAGE AROUND THE WORLD

Anthuriums growing in coconut husk in rural Jamaica: The roots of many plants grow very well in coconut fiber.

(PHOTO BY THE AUTHOR)

A beautiful but short-lived aechmea plant from the author's collection: The yellow, pink, white, and black flowers appear in March.

(PHOTO BY THE AUTHOR)

Snail damage is indicated by a series of holes in this calathea leaf. The snail feeds on the tubular unfurling leaf, and a curious pattern of holes develops.

(PHOTO BY THE AUTHOR)

The author with areca palm seed ready to harvest: The seeds have turned from green to orange.

(PHOTO BY BILL LEWIS)

Croton Banana on a Costa Rican stock farm, growing in full sun: This compact variety is excellent for small pots and dish gardens.

(PHOTO BY BILL LEWIS)

Cordyline growing as multi-trunked lattice in Amsterdam, Holland: Note the twist-ties holding the trunks.

(PHOTO BY MIKE RINCK)

Dieffenbachia Tropic Snow is a variety that originated as a mutation in a Fort Lauderdale nursery.

(PHOTO BY THE AUTHOR)

Mature foliage of *Dizygotheca elegantissima* is very different from its juvenile foliage. The leaves become much broader when the plant gets about 6 feet tall.

(PHOTO BY THE AUTHOR)

A *Dracaena fragrans* cv. Massangeana cane farm in the Dominican Republic: This cane will be ready to first-time harvest in two months.

(PHOTO BY THE AUTHOR)

The popular rabbit's-foot fern (*Davallia*) is not widely grown. Most production is for hanging plants.

(PHOTO BY ED CLAY)

Variegated *Ficus decora* growing in Jamaica: Today most rubber plants come from tissue culture.

(PHOTO BY THE AUTHOR)

The bright red and yellow flowers of the *Heliconia rostrata* can be used as cut flowers.

(PHOTO BY THE AUTHOR)

An interesting spiral hibiscus plant in Homestead, Florida: The stem was wound around a stake as the plant grew.

(PHOTO BY ED CLAY)

A mature *Phoenix reclinata* in an Orlando, Florida, theme park: This plant ultimately died from *Ganoderma*, a fungus disease.

(PHOTO BY THE AUTHOR)

Closeup of *Schefflera arboricola* flowers: The golden flowers are rarely seen in production nurseries, but are very common on mature specimens.

(PHOTO BY JOHN GATTI)

A South African grower proudly displays a nice Gold Capella *Schefflera arboricola*. Africa has some very fine nurseries.

(PHOTO BY THE AUTHOR)

These specimens of *Beaucarnia* or burro's-tail hanging baskets in El Salvador are many years old.

(PHOTO BY THE AUTHOR)

Sansevieria zeylanica stock plants in Costa Rica: Cuttings are grown in full sun, often needing little fertilizer or irrigation.

(PHOTO BY BILL LEWIS)

An old giant yucca in rural El Salvador: The yucca bloom is the national flower of El Salvador.

(PHOTO BY THE AUTHOR)

Yucca cane harvested in the wild by Guatemalan Indians, who sometimes use donkeys to transport cane from the fields.

(PHOTO BY THE AUTHOR)

Acalypha wilkesiana, the copper leaf, growing in full sun: These plants are for landscape production, though some are also grown for foliage.

(PHOTO BY THE AUTHOR)

Aspidistra, the cast-iron plant, with a slight chlorosis that is a mild case of magnesium deficiency.

(PHOTO BY THE AUTHOR)

A flat of Rex begonia liners: Begonias make good houseplants when provided with good light and well-drained soil.

(PHOTO BY THE AUTHOR)

These 10-inch fishtail palms *(Caryota mitis)* have six to eight seedlings per pot and are in 73% shade.

(PHOTO BY THE AUTHOR)

Saintpaulia ionantha, the African violet, blooms well in a north-facing window.

(PHOTO BY THE AUTHOR)

Strelitzia nicolai, the white bird-of-paradise, has white flowers that will develop to about twice the size of regular bird-of-paradise flowers.

(PHOTO BY THE AUTHOR)

is common in propagation. Treatments are similar as those for *Pseudomonas*. Dipping the cuttings from five to 15 minutes in a copper fungicide solution may reduce *Erwinia,* but the best control is to start with clean plant material.

A new vascular wilt of pothos caused by *Burkholderia solanacearum* has been found in Florida and Costa Rica. The bacterium causes plant wilt and discoloration of the intrnal stem tissue. There are as yet no chemical controls.

You may be a foliage grower, but what you should really be trying to grow are roots.

Insect and mite pests

Despite what other publications may indicate, two-spotted mites frequently attack pothos, especially in hot, dry weather. Mite feeding causes foliage speckling and yellowing. Numerous miticides are available, including Avid (abamectin), Pentac (dienochlor), and Talstar (bifenthrin). Mealybugs can be a big headache, as pothos vines provide many places for mealybugs to hide. When cottony masses from mealybugs are observed, sprays containing Cygon (dimethoate) or Dycarb (bendiocarb) are helpful, as is granular Di-syston (disulfoton). Root mealybugs may be treated with Diazinon drenches. Scale insects are not common on pothos, but when they occur, chemical controls are similar to those for mealybugs.

Thrips may cause scarring of leaf tissue, especially on one side of the leaf and stem, as thrips frequently feed inside the tube of the unfurling leaf. Sprays of Mavrik (tau-Fluvalinate) help control thrips. Finally, snails can be very destructive on pothos, creating numerous holes in the leaves and leaving squiggly droppings behind. Sprays with Grand Slam (methiocarb) are helpful, as are metaldehyde baits.

Disorders

Brown patches in the centers of pothos leaves frequently develop after exposure to low temperatures or after rapid temperature change. Brown patches in the variegated or white parts of pothos leaves occur under high temperatures. Leaf size is a good indicator of nitrogen fertility in pothos, assuming light levels are correct. Small leaves are usually related to lack of nitrogen or low light. Bird's-nest fungus *(Sphaerobolus)* appears similar to a scale insect. Captan sprays are beneficial.

Pothos is rather sensitive to Subdue (metalaxyl). When drenching with Subdue, use conservative rates. Subdue toxicity appears as marginal whitening or yellowing, especially of the older foliage. Do not drench pothos with Dycarb (bendiocarb) or Chipco 26019 (iprodione), or phytotoxicity may result. Manganese toxicity may cause a severe speckling of the older leaves, with the veins turning somewhat purplish. Older plants are most often affected, especially when medium pH is low. Liming helps, and be conservative in the fertilizer program. Pothos foliage becomes discolored after exposure to 2 ppm ethylene.

TRICKS

Many growers believe that color in pothos is largely genetic. That is true to some extent, but good nutrition can do a great deal to improve pothos variegation. Variegated pothos has various layers of cells in the leaf, some of which contain chlorophyll and some of which do not. The overlapping of these pigmented cell layers causes the attractive marbling. When plants are low in magnesium or if light is low, pothos spreads out its chlorophyll in order to catch more light, and variegation will be reduced.

Stock plants are virtually always deficient in magnesium because harvesting of cuttings removes a great deal of the plant's magnesium reserves. Magnesium nitrate sprays are very helpful in maintaining pothos color. Use 1 quart per 100 gallons (1 l per 400 l). However, in hot weather the grower may use a little less magnesium to reduce variegation and avoid burning of white leaf tissue.

Try not to let a pothos stock vine get more than 15 leaves, and when you make cuttings, cut back to where the remaining stock plant vine will still have four to five leaves. Pothos selections with larger leaves and thicker stems may be encouraged by training vines to grow upward. Use cuttings from those vines to establish stock with desired characteristics.

INTERIOR CARE

Pothos is one of the most durable, long-lived foliage plants for the interior environment. The normal room temperature range of 65 to 80F (18 to 27C) works fine. The plants do better with bright, indirect lighting. Pothos tolerates 50 f.c. (0.5 klux), but is more attractive with at least 150 to 250 f.c. (1.6 to 2.7

klux). Long-term storage under low light conditions results in reduced leaf size and less variegation.

Irrigate no more than once per week in winter, slightly more often in summer. Fertilize about three times per year, using 2 teaspoons of soluble 20-20-20 plus one-half teaspoon Epsom salts per gallon (3.8 l). If older leaves begin to turn canary yellow, the plants are either too wet or too dry. Checking soil moisture levels should help you make the correction.

REFERENCES

Chase, A.R. 1988. *Controlling Erwinia Cutting Rot of Marble Queen Pothos*. CFREC-Apopka Research Report RH-88-4: 1-5. Apopka: Central Florida Research and Education Center.

Chase, A.R., and R.T. Poole. 1990. *Effect of Variegation on Growth and Chilling Sensitivity of Marble Queen Pothos*. CFREC-Apopka Research Report RH-90-17: 1-2. Apopka: Central Florida Research and Education Center.

Chase, A.R., and R.T. Poole. 1991. Effect of Temperature on Rhizoctonia Root Rot of Pothos. *Nursery Digest* (January): 18-19.

Chase, A.R., and R.T. Poole. 1992. Effect of Potassium Rate, Temperature, and Light on Growth of Pothos. *Foliage Digest* 15 (5): 1-2.

Poole, R.T., L.S. Osborne, and A.R. Chase. 1984. Pothos. *Florida Nurseryman* (March): 51-52, 61.

Powell, C.C., and R. Rossetti. 1992. *The Healthy Indoor Plant*, 211. Columbus, Ohio: Rosewell Publishing, Inc.

Steinkamp, K., A.R. Chase, and R.T. Poole. 1992. *Pothos Production Overview*. CFREC-Apopka Research Report RH-92-16: 1-3. Apopka: Central Florida Research and Education Center.

Wang, Y.T. 1991. Maximizing Growth of Pothos Ivy. *Greenhouse Manager* (January): 74-78.

Wang, Y.T., and C.A. Boogher. 1988. Cultural Practices Affecting the Propagation of Golden Pothos. *Foliage Digest* 11 (9): 1-3.

FALSE ARALIA

(See Dizygotheca*)*

FATSIA

HABITAT

Fatsia is one of the few foliage plants that is not tropical in origin. The one species, *Fatsia japonica,* is native to USDA Zone 8 in Japan. It therefore prefers a milder, cooler climate and does not tolerate heat well. The name *Fatsia* is of Japanese origin, while *japonica* refers to being from Japan. The plant also does very well in Mediterranean and Californian types of climates.

Fatsias may reach 20 feet (6 m) in their native habitat, though as foliage plants and in the landscape, they are generally much smaller, usually 3 feet (0.m) or less. *Fatsia* is in the Aralia family, and it looks and behaves like other aralias. In fact, at one time *Fatsia* was called *Aralia sieboldii* in the trade. Common names include Japanese fatsia, Formosa rice tree, paper plant, and glossy-leaved paper plant. The common names don't really make a great deal of sense, so most foliage growers simply refer to the plant by its genus name.

Fig. 22 *Fatsia japonica* from Illinois. (Courtesy of John Gatti)

USES

Most fatsias are produced as single, freestanding potted plants. The most common pot sizes are 6 to 10 inches (15 to 25 cm). Three plants per pot are occasionally used in larger pots, but with the large leaf size, single plants tend to be preferred. The petioles are about a foot (30 cm) long, and the medium to deep green leaves have five to nine deep lobes. Small, white flowers occasionally appear, though they are not particularly showy.

In addition to being floor plants, fatsias make excellent landscape plants in mild climates. They do very well with north side exposure, as they are rather cold tolerant and don't care for high temperatures. Fatsias are frequently used in the landscape to create an oriental effect.

VARIETIES

In addition to the common variety, there is a Variegata cultivar that has white blotches near the leaf tips. An attractive cultivar of French origin known as Moseri is similar to the basic fatsia but is more compact and has broad leaves. Moseri is also more vigorous than its parent. A curious variation is ×️ *Fatshedera lizei*. This is an intergeneric cross between *Fatsia* and *Hedera*, the English ivy. The leaves are similar to the fatsia's, though they are only about 6 inches (15 cm) across, and they grow on ivylike stems. A variegated form of *Fatshedera* also exists.

PROPAGATION

Propagation from seed, cuttings, root cuttings, and air layers have been successful. When growing from seed, it is best to separate the seeds from the fruits, as germination is much better. Sow them barely covered, preferably at a soil temperature of 80F (26.7C). Germination typically runs in the 60% range.

Propagation from cuttings is primarily done from the variegated forms, which generally do not come true from seed. The best cuttings are taken from crisp, slightly woody tissue in summer. A light rate of rooting hormone may be beneficial, and the cuttings need to be protected from high air temperatures, although they like bottom heat. Mist conservatively or use a propagation tent.

CULTURE

Some references say to grow *Fatsia* under high light, at 4,000 to 6,000 f.c. (43.2 to 64.8 klux). This works in cool, dry climates, but in warmer, more humid environments, you must grow fatsias much darker, typically at around 73%

shade, or light levels in the range of 1,000 f.c. (10.8 klux). As with some dracaenas, you are shading for temperature control more than for light reduction.

Because fatsias have relatively weak roots and are susceptible to root disease, it is important to use a mix with a somewhat lower moisture-holding capacity than usual. The mix must also have excellent drainage and be a very open, loose mix. Mixtures of 2:1:1 peat, bark, and cypress shavings have been successful. Try to keep the peat at 50% or less so that the medium does not retain excessive moisture.

Due to their temperate nature, fatsias should be grown at cooler temperatures, which can be an advantage for the northern grower. Common daytime greenhouse temperatures for good fatsia production generally range from 62 to 70F (17 to 21C), but 50F (10C) is fine for *Fatsia*. Night temperatures can easily be 45 to 50F (7 to 10C). Colder temperatures slow their growth, but they do not tend to suffer cold injury until temperatures are close to freezing. Fatsias are ideal for high-altitude, low-humidity situations.

NUTRITION

Being somewhat slow growing, fatsias are also relatively light feeders. Just 6 grams of slow-release, coated three-month 18-6-8 or 19-6-12 fertilizer per 5-inch (12.5-cm) pot is adequate. Like many other slow-growing varieties, they like a gentle yet steady supply of nutrients. They are incapable of flushing rapidly after high rates of fertilization. The relatively weak root system also makes phosphorus fairly important. To help stimulate better root growth, I like to either incorporate superphosphate in the medium or drench at potting with a high-phosphate starter fertilizer.

Fatsias have relatively few nutritional disorders. Lack of sulfur tends to make them light in color. In wet media they may show paleness in the new foliage, due to lack of iron. When overall fertility is low, leaf expansion and color are poor. High soluble salts can cause leaf tip burn at times. Fertilizing lightly should enable you to avoid soluble salt problems.

DISEASES

For many plants in the family Araliaceae, *Alternaria panax* is a common leaf spot and blight problem. It is much more severe under high light conditions, so producing *Fatsia* under darker conditions greatly reduces incidence of *Alternaria*. Try to keep the foliage dry, and spray with either Dithane (mancozeb) or Daconil (chlorothalonil). Anthracnose, caused by the fungus

Colletotrichum, is also a brown leaf spot, though the lesions are generally smaller. This fungus is less affected by light. Control options are similar to those for *Alternaria*, though Zyban (thiophanate methyl plus mancozeb) is also used.

When fatsias are overwatered, root rots from *Rhizoctonia*, *Pythium*, or *Phytophthora* can occur. Moisture control and drainage are, of course, critical in such a situation. Fungicide drenches with Banrot (ethazol plus thiophanate methyl) are generally helpful. Southern blight can also infect plants during hot weather. The fungus's tan to white fruiting bodies and white threads (mycelia) are highly visible. Discard affected plants and treat the remainder with either Terraclor (PCNB) or the insecticide Dursban (chlorpyrifos).

INSECT AND MITE PESTS

While fatsias are not particularly prone to pests, spider mites can cause trouble. The problem is compounded by the fact that the orientation of the foliage makes it difficult to spray the undersides of the leaves. The translaminar action of Avid (abamectin) is very helpful in this situation. Thrips cause leaf scarring and distortion and occasionally scarring of petiole tissue. Sprays of Mavrik (tau-Fluvalinate) or Orthene (acephate) are useful, as is Avid at high rates, but be careful spraying Orthene on fatsias in hot weather. Scale insects and mealybugs attack fatsias, but only very infrequently.

 When soluble salts are high and you can't leach, watering with sugar water stimulates soil microbe activity to reduce soluble salts.

DISORDERS

Heat breakdown is the most common nursery disorder with fatsias. The plants try to grow, but they rapidly become unattractive, and mortality can be high. Overwatering even briefly can greatly reduce plant quality. Keeping light low helps plant quality in hot climates, but it can also make it difficult to dry plants out if they become too wet.

Pale color is frequently the result of lack of sulfur in the fertilizer program. Spider mite activity can cause speckling of foliage and may contribute to marginal leaf burn and dieback. Even mildly elevated soluble salts can cause burn and leaf drop, as well as root injury.

TRICKS

While fatsias appear relatively delicate, they can take a little bit of salt. Growers with somewhat salty irrigation water may still be able to grow fatsias, if they use care. The growth regulator B-nine (daminozide) is an excellent tool for keeping fatsias compact and improving their leaf color. One or two sprays at 2,000 ppm give excellent results.

Grow fatsias on the dry side, and be sure to let the medium dry out between waterings. It is best to have them in a covered greenhouse to protect them from excess rain. This is just a hunch, but I strongly suspect that inoculation with mycorrhizal fungi will result in much better root health and vigor. Picking off the flower buds when they form will help maintain larger leaves on the plant.

INTERIOR CARE

Japanese fatsias do best in cooler interior environments. The preferred temperature range is 62 to 70F (17 to 21C). They tolerate relatively cool nights without incident, even if temperatures reach 45F (7C). Indirect sunlight is helpful, as the plants tend to break down below 100 f.c. (1.1 klux); 150 to 250 f.c. (1.6 to 2.7 klux) is preferred.

As in the nursery, these plants should be maintained relatively dry indoors. If soil moisture is either too wet or too dry, leaf drop and loss of vigor rapidly occur. Plants stretch and become unattractive fairly quickly if light is too low. Make sure fatsias have decent root systems before you purchase them to help ensure good interior performance.

REFERENCES

Bailey, L.H., and E.Z. Bailey. 1976. *Hortus Third*, 471. New York: Macmillan.

Blessington, T.M., and P.C. Collins. 1993. *Foliage Plants: Prolonging Quality*, 117-119. Batavia, Ill.: Ball Publishing.

Chase, A.R. 1986. *Factors Affecting Growth of Fatsia*. AREC-Apopka Research Report RH-86-10: 1-2. Apopka: Agricultural Research and Education Center.

Dirr, M.A., and C.W. Heuser. 1987. *The Reference Manual of Woody Plant Propagation*, 124. Athens, Ga.: Varsity Press Inc.

Watkins, J.V., and T.J. Sheehan. 1975. *Florida Landscape Plants*, 306. Gainesville: The University Presses of Florida.

FERNS

HABITAT

The foliage industry in Florida more or less began in 1912 to 1914 with the first serious production of Boston ferns. However, in this book I will deal with some of the major classes of cultivated ferns. You will not find asparagus here because it is not a fern; it was covered earlier. I have also not discussed *Rumohra*, the leatherleaf fern, because it is generally grown for cut foliage and rarely as a houseplant. I will cover the genera *Adiantum, Asplenium, Davallia, Nephrolepis, Platycerium,* and *Pteris*. Alas, it is a challenge to write about so many different ferns in one section without it becoming a garbled mess, but I will try.

The 10,000 species of fern are seedless and flowerless plants whose true roots emanate from a rhizome, or from an underground stem. Ferns grow naturally all over the world, even in the Arctic. Some grow as large as 80 feet (24 m)! *Adiantum*, the maidenhair fern, comes from temperate North America as well as tropical America and East Asia. With its temperate roots, it tends to go dormant in winter. The staghorn

Fig. 23 *Nephrolepis exaltata*, the Boston fern, growing wild in its native habitat of southern Florida. (Photo by the author)

ferns *(Platycerium)* come from Queensland, Australia, and can therefore tolerate a brief freeze. *Pteris* ferns hail from the Himalayas to Ceylon and the South Pacific. They like cool nights and can take higher light than many ferns.

Asplenium species, the bird's-nest ferns, are epiphytes from tropical Asia and Polynesia. They therefore need warmer temperatures. Boston ferns (*Nephrolepsis*) are native to many tropical regions and thus prefer warm, humid, dark conditions. *Davallia*, the rabbit's-foot fern, comes from Fiji and the Malay archipelago. Because of its tropical origins, it tends to go deciduous below 60F (15.6C).

USES

Fern growers have many options in terms of varieties and types of production. Four- and 6-inch (10- and 15-cm) azalea pots are commonly used for ferns, as are hanging baskets ranging from 6 inches to 12 inches (15 to 30 cm). Both clay and plastic pots are acceptable, and hanging baskets do especially well in unlined containers made of wire and osmunda.

Some growers produce 2- and 3-inch potted liners for other growers to step up. Epiphytic ferns, such as the staghorns, are grown on plaques with sphagnum moss or in wire baskets.

 Metaldehyde baits don't directly kill snails, instead baits intoxicate snails and they die of desiccation.

VARIETIES

Popular maidenhair varieties include the Delta maidenhair, *Adiantum raddianum,* and the dwarf known as Baby's Tears. Ocean Spray and Pacific Maid are popular for baskets. *Asplenium* cultivars include Antiquum, known as the Japanese bird's-nest fern. *A. crispafolium* is sometimes called the lasagne fern, as its wavy leaves look like lasagne noodles. The most common *Davallia* varieties include rabbit's-foot fern, *D. fejeensis* cv. Plumosa, as well as the squirrel's-foot fern and the ball fern.

Among the many Boston fern cultivars are Bostoniensis, Boston Compacta, Kimberly Queen, Florida Ruffle, Rooseveltii, and Dallas. I don't have space to describe all of these, but all are variations of the classic Boston fern. In the trade, *Platycerium* is usually *P. bifurcatum.* The variety Netherlands is darker green, with broader, less lobed leaves. *P. hillii* is the elegant elk's-horn fern. Many *Pteris* ferns exist, including *P. ensiformis* cv. Victoriae, which has small, silver and green leaves.

PROPAGATION

Ferns may be propagated by various methods, including division of stolons, root clumps, and plantlets, as well as from spores and tissue culture. Most of the newer hybrids are produced from tissue culture, frequently in 72-plug trays. With spores that are usually not viable, Boston ferns are produced from stolons, or from tissue culture with the improved varieties. Spores of *P. enformis* cv. Victoriae sprout in about three weeks in acidic sedge peat, preferably under either mist or plastic cover. Staghorn ferns are often propagated by removing the offshoots (pups) with a sharp knife. Try to include some roots with the pup, if you can. Rabbit's-foot ferns may be propagated by breaking off a foot, laying it on moist soil, and covering with plastic until it sprouts.

CULTURE

Maidenhair ferns are usually grown between 1,200 and 1,800 f.c. (12.9 and 19.3 klux), with minimum night temperatures of 60 to 65F (16 to 18C). They like at least 50% humidity and tolerate few pesticides. They go dormant in the winter, so it is a good idea to let them rest or to limit your season of production to the warmer months. Bird's-nest fern is usually produced between 1,500 to 2,000 f.c. (16.2 to 21.6 klux). It likes a shallow pot, as it has a fairly weak root system. The preferred daytime temperature is 70 to 90F (21.1 to 32.2C), with a minimum of 60F (16C) at night. Don't plant it deep, and try to keep the crown dry. *Davallia* is best produced at 60F (16C) or above. Instead of a traditional container, it does better with a basket or carved container of osmunda. Don't try to grow it in regular potting soil.

Two thousand f.c. (21.5 klux) are ideal for Boston ferns, with the normal range being 1,500 to 3,000 f.c. (16.2 to 32.4 klux). A mix of 60% sphagnum and 40% perlite or polystyrene beads will perform well. Initial pH should be in the 5 to 5.5 range, although the operating pH is usually 4 to 5. Growth may be reduced 25 to 50% if the pH rises to between 6.2 and 7. Best daytime temperatures are 65 to 95F (18 to 35C). However, research shows that the ideal day and night temperature is not far from 72F (22.2C). The grower usually plants one liner in a 4-inch (10-cm) pot, two liners in a 6-inch (15-cm) pot, and four liners in an 8-inch (20-cm) pot.

Staghorn ferns are best produced in sphagnum moss with a pH of 5 to 6. The usual minimum night temperature is 50F (10C), though staghorns can tolerate a brief freeze. *Pteris* ferns need good drainage, a minimum night temperature of 55F (12.8C), and light levels similar to those for maidenhair ferns.

In general, don't overpot ferns because their limited root systems often don't have the capacity to utilize a large container quickly.

NUTRITION

Ferns typically have low trace element requirements. Most growers use a 3-1-2 or 2-1-2 ratio of fertilizer, with liquid fertilization being the most popular. Most growers use 100 to 200 ppm nitrogen about once a week. Staghorns usually receive 100 ppm nitrogen about once a week. Keep the feed rates toward the low end in winter. It is best to rinse the liquid feed off the foliage in order to reduce the chance of burn.

You will usually find that incorporating trace elements into the potting medium is not necessary as long as there are some trace amounts of micronutrients in your liquid fertilizer. Because ferns like to grow at lower pH values, trace element toxicities can occur if you use trace elements aggressively. Ferns generally like sulfur; in many fern varieties, a lack of sulfur will cause brittle, weak growth, with poorly formed leaflets. Remember that most soluble fertilizers do not contain sulfur. Magnesium sulfate is a convenient, inexpensive sulfur source for ferns.

DISEASES

Rhizoctonia aerial blight is one of the most common fern diseases. It normally occurs in summer conditions of high humidity with temperature ranges of between 75 and 95F (23.9 to 35C). You rarely see this disease in winter. A rather severe blight, it can happen quickly, often in less than a week. The various parts of the foliage develop large brown to black blotches, especially in the internal parts of the plant, where drying conditions are reduced. The fungus often emanates from the soil where moisture is prevalent. Foliar sprays of Daconil (chlorothalonil) or Terraguard (triflumizole) are effective.

Pythium root rot, the second most common fern disease, causes stunting, wilting, and yellowing of fronds. The cortex, or shell, of the root tends to slough off, leaving the internal string of the root exposed. Parts of the roots may appear gray and water soaked instead of normal and healthy. Dry the plants down when fighting *Pythium,* and drench with either Truban (etridiazole) or Subdue (metalaxyl).

Many ferns are bothered by lesion nematode *(Pratylenchus)*. The symptoms are very similar to root rot, and in most of the United States there are currently no chemical controls. The best answer right now is to start with and maintain clean plants, discarding any infested ones. *Asplenium* is bothered by a foliar nematode, *Aphelenchoides*. Large, brown to black, water-soaked lesions develop. The nematodes have trouble spreading across the midvein, so the spots are somewhat angular and tend to stop at the midvein. There are no nematicides available for most of the United States, so remove and discard infected leaves with sterilized clippers.

The bird's-nest ferns also suffer from two bacterial diseases caused by *Pseudomonas cichorii* and *P. gladioli*. Small, clear, water-soaked spots form all over the plants. Try to keep the leaves dry; sprays with Phyton 27 (picro cupric ammonium formate) may be helpful. As with all ferns and all chemicals, however, check for phytotoxicity first because ferns are very sensitive to chemicals.

INSECT AND MITE PESTS

Caterpillars frequently cause chewing injury on many types of ferns. The injury is usually obvious, as are the droppings. Sprays of Dipel *(Bacillus thuringiensis)* are helpful, as are Orthene (acephate), Sevin (carbaryl), and others. Mealybugs are not a common problem in ferns, fortunately, but they do occur. Their cottony masses are usually found on the undersides of the lower fronds, sometimes in the roots. Sprays of Mavrik (tau-Fluvalinate) may be helpful. Root mealybugs may be controlled with Diazinon. If you have just a few mealybugs, it might be better to discard the infested plants.

Snow scale, white fern scale, and brown scale are found on ferns. Dycarb (bendiocarb) may be useful, but it is also phytotoxic to some fern varieties, including some of the Boston ferns. Thrips occasionally distort some fern varieties, causing gnarled foliage to emerge. Sprays of Mavrik (tau-Fluvalinate) or Orthene (acephate) control them, especially when combined with a wetting agent. Fungus gnats are a problem once in a while. Gnatrol *(Bacillus thuringiensis)* can be helpful, as are good water management and yellow sticky cards.

DISORDERS

Ferns tend to turn gray when excessively dry. When this happens, irrigate very thoroughly to rehydrate the media. Weak frond growth is usually an indication of low light, while pale fronds suggest that light levels are too high. Burn of the leaves and runners indicates elevated soluble salts.

Agrimycin 17 (streptomycin sulfate) and Ornalin (vinclozolin) may be phytotoxic on a number of fern varieties. High nitrogen fertility causes bird's-nest ferns to send distorted, dented, or multilobed leaves. Reduce fertility if this happens. Necrotic leaf edges may indicate copper toxicity if copper fungicides are being used. Staghorn ferns are rather sensitive to excess manganese, which causes discoloration of the older foliage.

TRICKS

When irrigating ferns, it is best to let the water run through the medium to avoid accumulation of soluble salts. Don't let ferns go significantly dry very often, or you will see performance problems. Maidenhair ferns like extra calcium in the medium for good growth quality. Most ferns require low amounts of lime in the soil but benefit from gypsum in the potting soil. They seem to like the additional calcium and, especially, sulfur. Use about 3 pounds per cubic yard (1.8 kg per 3m).

Boston ferns throw more fronds in the winter in northern greenhouses if mum lighting is used to alter day length. Most ferns do better indoors if they are first placed in 750 f.c. (8 klux) for a week. Increasing temperatures means longer fronds in Dallas ferns.

INTERIOR CARE

The main trick with ferns, as with many foliage plants, is to maintain steady, consistent moisture levels, avoiding extremes in either direction. Don't permit any standing water to remain in saucers underneath ferns. Apply liquid fertilizer very sparingly every three months or so.

Ferns in general don't like dry heat. They actually prefer a moist, cool, slightly clammy environment. Many varieties, especially maidenhair ferns, like at least 50% humidity.

Minimum light levels for most ferns are 75 to 150 f.c. (0.8 to 1.6 klux). A better light level is around 250 f.c. (2.7 klux), but that may be difficult to achieve in many situations.

Maidenhair ferns do well with light from a north window or artificial lighting in the typical fern range. These plants go dormant in the winter, so water very sparingly and let them rest. Care of bird's-nest ferns is similar to that of maidenhair ferns. Boston ferns do well in slightly warmer interior environments, preferably from 60 to 75F (16 to 24C). As do other ferns, they appreciate good humidity levels and reasonable soil moisture. Staghorn ferns tolerate cool indoor environments.

REFERENCES

Blessington, T.M., and P.C. Collins. 1993. *Foliage Plants: Prolonging Quality,* 10-21, 37-39, 147-148, 170-171. Batavia, Ill.: Ball Publishing.

Chase, A.R., and C.A. Conover. 1988. Temperature and Potting Medium Composition Affect Rhizoctonia Aerial Blight of Boston Fern. *Foliage Digest* 11 (2): 1-3.

Chase, A.R., R.T. Poole, and L.S. Osborne. 1984. *Bird's-Nest Fern.* AREC-A Foliage Plant Research Note RH-1984-I: 1-3. Apopka: Agricultural Research and Education Center.

Conover, C.A., L.S. Osborne, and A.R. Chase. 1987. Boston Ferns. *Foliage Digest* 10 (3): 1-4.

Davenport, E. 1977. *Ferns for Modern Living,* 14-19, 24-25, 34-36, 40-47, 52-53, 61-67. Kalamazoo, Mich.: Merchants Publishing Company.

Erwin, J. 1990. Culture notes. Boston fern, Family: Polypodiaceae, Genus and species: *Nephrolepis exaltata* 'Bostoniensis'. *GrowerTalks* (February): 12.

Hipp, B.W., and D. Morgan. 1980. Influence of Medium pH on Growth of Roosevelt Ferns. *Hortscience* 15 (2): 196.

Poole, R.T., and C.A. Conover. 1983. *Fertilization of Bird's-Nest Fern.* ARC-A Research Report RH-83-18: 1-5.Apopka: Agricultural Research Center.

Poole, R.T., and C.A. Conover. 1984. *Fertilization of Staghorn Fern.* ARC-A Research Report RH-1984-5: 1-4. Apopka: Agricultural Research Center.

Soukoup, J., and F. Ohme. 1984. The Finest Ferns. *Florist's Review* (March 22): 22-25.

FICUS BENJAMINA AND FICUS RETUSA

Weeping Fig

HABITAT

There are over 800 species and 2,000 varieties of *Ficus*, most of which are native to the Old World tropics. The genus *Ficus*, whose name is the ancient Latin for "fig," is a branch of Moraceae, the mulberry family. *Ficus benjamina*, commonly called weeping fig, originally comes from India, Southeast Asia, and northern Australia. *Ficus retusa nitida*, known in the trade simply as *Ficus nitida*, though commonly called the Indian laurel or Cuban laurel, hails from the Malay Peninsula to Borneo.

Ficus benjamina has drooping, trailing branches, whereas the *Ficus nitida* branches are more upright and spreading. In their native habitats these plants may be found growing in the dense forests, either in full sun or under the heavy natural shade of their neighbors. *F. benjamina*, especially, comes from an area where there are distinct wet and dry seasons. Ficus trees usually shed

Fig. 24 *Ficus benjamina* grown as a standard. (Photo by the author)

leaves at the onset of the dry season in order to survive until the rains return. These factors help explain the tolerance of ficus plants to various light levels as well as its sensitivity to moisture stress.

Uses

Ficus trees were introduced to the foliage trade in the late 1950s. They are probably used in more different ways than any other foliage plant. Pot sizes range from 4 inches (10 cm) to huge tubs of over 200 gallons (757 l), and everything in between. *Ficus* is produced as bushes, standard trees, and multiple-trunk trees. Trunks can be braided, spiraled, poodled, knotted, or grown as bonsai. *F. benjamina* and *F. nitida* are frequently seen in tropical landscapes, either as large trees or closely trimmed hedges of varying heights.

Varieties

Several varieties have been developed over the years, with many coming from Denmark and the Netherlands. *F. benjamina* cultivars include Spire, which has a very narrow, columnar habit. Foliole, with small leaves and a very drooping appearance, is useful in small containers and for bonsai. Variegated cultivars include Jacqueline and the very similar Golden King. These have grayish leaves with green centers, and irregular ivory and light gray-green borders. Spearmint is similar to Jacqueline, though less gray in appearance, and its leaves are not as broad. Wintergreen is a nonvariegated variety that has darker foliage, especially the new growth.

F. *nitida* varieties include Green Gem, which is more vigorous than the regular *F. nitida* and has more symmetrical growth. Hawaii is a highly variegated version of *F. nitida*. The variety Nuda has been attributed to both *F. benjamina* and *F. nitida* parentage in the past, though it is probably actually a cultivar of *Ficus stricta*. It came from Texas in the late 1970s and has large leaves, a lighter trunk color, and a very weeping habit.

Propagation

Most ficuses root fairly easily from cuttings, usually under frequent mist (five seconds every 10 minutes) or fog. Rooting hormones are generally not required. Bottom heat may be helpful for rooting in cool situations. Shoot for a soil temperature of about 82F (28C). Stock plants in many parts of the world produce seeds which are not viable.

For growing standard trees and braids, air layers are commonly produced from stock hedges. Lower leaves are removed from straight new shoots. Bark is removed from the basal portion of the stem, which is wrapped in sphagnum and then aluminum foil or polyethylene. Cuttings root in three weeks or so, whereas air layers take longer to root well. Also, many of the newer cultivars, including the variegated types, are now produced primarily from tissue culture.

Waterlogged soil generates ethylene.

CULTURE

Ficuses respond well to regular, thorough irrigation. Research shows growth is significantly reduced as watering frequency decreases. Soil mixes need to be reasonably well drained, however, or root rot can develop. Blends of peat and either bark or wood chips are common, typically with 10 to 20% sand to help keep the plant from blowing over in the wind.

Production light levels usually range from 3,500 to 5,000 f.c. (37.7 to 53.8 klux). This is 50 to 70% shade in tropical areas. Many growers like to grow ficuses in full sun for a while at first to develop thick, sturdy trunks, although this can lead to leaf drop later in the interior environment. The tissue culture varieties are normally produced entirely under shade. When grown in the shade, ficus leaves are larger and darker than sun-grown leaves but also thinner. The number of stomates under shade is larger, and they have more chlorophyll in the upper leaf cells. The longer the grower keeps ficuses under shade, the better the plants in the interior, especially large specimens.

If the soil temperature goes above 95F (35C), roots and plant height may be reduced. Soil temperature can reach 105F (40.5C) during midday under full sun in the tropics, so midday waterings may be needed to help reduce soil temperature. Plants are damaged as temperatures approach freezing.

NUTRITION

Because of their requirement for frequent waterings, ficuses also need fairly aggressive fertilization. Granular fertilizers of approximately 3-1-2 ratio, such as 12-6-8 or 7-4-5, are popular, frequently applied every $1^1/_2$ to two months. Constant liquid feed of 200 ppm nitrogen from 20-10-20 is also common, especially for smaller containers. Slow-release fertilizers may be top-dressed or incorporated into the mix, but use the higher rates, or else the heavy nutrient demands may not be met.

F. benjamina, especially, has fairly high magnesium requirements, though the lack of magnesium may show up in new as well as older leaves. Manganese deficiency causes severe chlorosis of new leaves when medium pH is high. Ficuses like calcium, and when grown with hard irrigation water, they tend to be stronger and more compact than when grown with soft water. Growers with low-calcium irrigation water may benefit from foliar calcium sprays. Lack of boron can make ficus plants rather weak looking, even when light levels are satisfactory.

TABLE 15 Leaf analysis rating standards for *Ficus retusa nitida*

Nutrient (%)	Very low	Low	Medium	High	Very high
Nitrogen	<1.25	1.25–1.49	1.50–2.50	2.51–4.00	>4.00
Sulfur	<0.12	0.12–0.19	0.20–0.50	0.51–1.00	>1.00
Phosphorus	<0.08	0.08–0.09	0.10–0.50	0.51–1.00	>1.00
Potassium	<0.75	0.75–0.99	1.00–2.50	2.51–4.00	>4.00
Magnesium	<0.20	0.20–0.24	0.25–0.50	0.51–1.00	>1.00
Calcium	<0.40	0.40–0.69	0.70–2.50	2.51–3.50	>3.50
Sodium			<0.21	0.21–0.50	>0.50
(ppm)					
Iron	<30	30–49	50–300	301–1000	>1000
Aluminum			<251	251–2000	>2000
Manganese	<15	15–19	20–200	201–1000	>1000
Boron	<15	15–24	25–50	51–75	>75
Copper	<4	4–5	6–100	101–500	>500
Zinc	<11	11–14	15–200	201–1000	>1000

Sources: Institute of Food and Agricultural Sciences, Apopka, Florida; Dr. Benjamin Wolf, Fort Lauderdale, Florida.
Notes: Common name: Cuban laurel. Sample of most recent fully mature leaves, no petioles.

DISEASES

The most troublesome disease of *F. benjamina* is *Phomopsis.* The fungus attacks other species of *Ficus,* as well, though usually less aggressively. *Phomopsis* causes a twig dieback disease that can be very severe in interior situations. It is usually a weak pathogen in nurseries, but the fungus proliferates on dead twigs in interior environments, especially when plants are exposed to water stress. Combat the disease by avoiding water stress, removing dead twigs with sterilized clippers, and spraying where appropriate with a thiophanate methyl fungicide. It is important to sterilize clippers with either rubbing alcohol or a bleach solution, as *Phomopsis* is frequently mechanically transmitted.

When soil is excessively moist, root rot diseases caused by *Pythium, Fusarium,* or *Rhizoctonia* can develop. Usually, reducing soil moisture enables the plants to recover, but they may require fungicide drenches in extreme situations. One frequently seen bacterial disease is a leaf spot caused by *Xanthomonas campestris* pv. *fici.* The disease attacks most ficus cultivars, especially Nuda. Increased fertility rates have been shown to reduce *Xanthomonas* leaf spot in *F. benjamina*. Sprays with Dithane (mancozeb) plus a copper fungicide should provide good control. Keeping foliage dry as much as possible also helps.

INSECT AND MITE PESTS

Ficuses are not particularly prone to insect problems. They are occasionally affected by infestations of mealybugs or any of several different scale insects. In addition to oil sprays and various biological controls, sprays of Dycarb (bendiocarb), Cygon (dimethoate), or Mavrik (tau-Fluvalinate) are effective. Granular Di-syston (disulfoton) is also very good for nurseries and stock farms.

The biggest problem for *F. nitida* is the Cuban laurel thrips. Numerous black thrips proliferate and feed inside curled-up, new foliage. The leaves become distorted, with numerous brown spots and gray feeding scars. A few *F. nitida* strains are resistant to thrips, but most may require occasional sprays of Orthene (acephate) or Mavrik (tau-Fluvalinate) in the nursery.

F. benjamina is frequently bothered by low-level infestations of red spider mites. The mites are not a big problem in nurseries, but many growers fail to realize that the mite infestations are there. When the plant is shipped to an interior environment, however, the mite problem becomes more acute. Growers should monitor their ficuses for mites and treat with either predator mites or one of the many registered miticides.

DISORDERS

Leaf drop is the most common problem on *Ficus*, especially when a plant is moved into an interior or otherwise shaded environment. Moisture stress and low light contribute significantly to leaf drop, and abscisic acid levels increase as ficus plants go into interior environments. What really happens is that ficuses generate internal ethylene when exposed to moisture stress. The plants think it is the dry season coming, and leaves are shed. Exposure to ethylene causes leaf loss, as does low-level exposure to mercury from paints. Moisture-stressed ficuses tend to drop yellow leaves, whereas green leaves are shed when exposed to ethylene, low-light stress, or mercury.

If heavy rains occur right after application of granular fertilizers high in sulfur, rapid and severe leaf drop may occur. I suspect this is caused by sulfide toxicity, generated by anaerobic soil microbes. The plants usually quickly recover completely. Ficus bushes grown close together may severely shed lower leaves after heavy rains, due to ethylene generation from waterlogged soil. Algae may grow on trunks in wet environments. During heavy rains, production of aerial roots may increase. Internode length and a droopy appearance may develop if plants are grown under low light or without adequate calcium.

TRICKS

Large, field-grown tree specimens may perform better in interiors if more potting soil, and less field soil, is used in the containers. Keeping young standards and braids pot bound helps develop heavy stem caliper. Providing adequate space and regular shearing also promote thicker stems. Closer plant spacing may help reduce soil temperature in full-sun situations. Withholding irrigation tends to slow growth and may contribute to leaf drop.

When producing a braid, use four stems of equal length and thickness. When the braid is completed, shake it to ensure that it will hold together. When *F. benjamina* is growing extremely rapidly, the new leaves come out red. Ficus plants manufacture special shade proteins, and shade-grown plants have a lower light compensation point, which means they simply need less light to survive.

INTERIOR CARE

Light and moisture are critical for interior ficuses. Light level should be at least 150 to 250 f.c. (1.6 to 2.7 klux). Bright indirect light is desirable, and the more light, the better. Moisture status needs to be monitored regularly, and moisture stress should be avoided at all costs. Preferred temperatures are the normal interior range of 65 to 80F (18 to 27C). *Ficus* can tolerate brief spells of cool weather without incident, however.

Fertilize ficuses about every three months in the interior, and include magnesium and manganese. A useful mixture would be 2 teaspoons of soluble 20-10-20 per gallon, along with 1 teaspoon magnesium sulfate and one-fourth teaspoon manganese sulfate.

Prune any observed dead branches regularly. Sterilize your cutting tools frequently with alcohol or bleach solutions in order to reduce spread of disease. Watch carefully for leaf drop and twig dieback symptoms. *F. nitida* generally has fewer leaf drop problems than *F. benjamina*.

REFERENCES

Anderson, R.G., and J.R. Hartman. 1981. *Phomopsis* Twig Blight on Weeping Fig Indoors: A Case Study. Publication unknown: 5-7.

Chase, A.R. 1990. *Effect of Nitrogen and Potassium on Growth of* Ficus benjamina *and Severity of* Xanthomonas *Leaf Spot*. CFREC-Apopka Research Report R-90-3. Apopka: Central Florida Research and Education Center.

Conover, C.A., and R.T. Poole. 1990. *Acclimitazation Revisited*. CFREC-Apopka Research Report RD-90-2. Apopka: Central Florida Research and Education Center.

Graves, W.R., and R.J. Gladon. 1985. Ficus and Leaf Drop: An Update. *Interior Landscape Industry* (March): 61-64.

Henley, R. 1991. An Overview of Ficus for Interior Landscapes. *Greenhouse Manager* (January): 62-68.

Johnson, C.R., T.A. Nell, J.N. Joiner, and J.K. Krantz. 1979. Effects of Light Intensity and Potassium on Leaf Stomatal Activity of *Ficus benjamina* L. *Hortscience* 14 (3): 277-278.

Johnson, C.R., T.A. Nell, S.E. Rosenbaum, and J.A. Lauritis. 1982. Influence of Light Intensity and Drought Stress on *Ficus benjamina* L. *J. Amer. Soc. Hort. Sci.* 107 (2): 252-255.

Peterson, J.C., J.N. Sacalis, and D.D. Durkin. 1981. *Ficus benjamina:* Avoid Water Stress to Prevent Leaf Shedding. *Florist's Review* (January): 10-11, 37-38.

Poole, R.T., and C.A. Conover. 1991. Mercury Toxicity to *Ficus* spp. *Foliage Digest* (July): 7.

Wang, Y.T. 1988. Influence of Light and Heated Medium of Rooting and Shoot Growth of Two Foliage Plant Species. *Hortscience* 23 (2): 346-347.

FICUS LYRATA:
Fiddle-Leaf Fig
FICUS ELASTICA:
Rubber Plant
FICUS MACLELLANDII:
Alii

HABITAT

These three *Ficus* species (family Moraceae) are popular as larger potted bush or tree forms for interior use. They are tropical forest trees, which makes them fond of good light levels. They prefer the moist conditions of their native environments and tend to have very aggressive root systems. Their tropical forest origins make these trees significantly more tolerant of heat than cold.

Ficus lyrata (not *F. pandurata*) is commonly called the fiddle-leaf fig due to its violin-shaped leaves.

Fig. 25 A full *Ficus lyrata*, grown in 63% shade. (Photo by the author)

The large, veiny, emerald green leaves are about 15 inches (37.5 cm) long. The tree reaches a height of 40 feet (12.3 m) in its native habitat.

Ficus elastica, commonly known as the India rubber tree or simply the rubber tree, originally comes from southern Asia from Nepal to Burma. The thick, succulent leaves average 12 inches (30 cm) long and come in several colors. In the tropics, rubber trees reach 40 feet (12.3 m), with a widespread canopy.

Ficus maclellandii is commonly known in the trade by the varietal name Alii, which means "king" in Hawaiian. Alii is smaller and slower growing than the other two species listed here. The olive green leaves are long and narrow, generally hanging somewhat downward.

USES

The relatively large leaves of these three ficus varieties make them preferable as larger specimens. *F. lyrata* is commonly grown in bush form in 6- and 10-inch (15- and 25-cm) containers, usually from tissue culture. Standard tree forms are also produced in 10-inch and larger containers. *F. lyrata* has been popular in the trade for about the last 15 years.

Cultivars of *F. elastica* have been popular houseplants since the turn of the century. Some 6-inch production of single cuttings or air layers exists, but most of the production today, at least in the United States, is in 10-inch containers from tissue culture liners. Standard and multitrunked trees are also produced in larger containers, usually from cuttings or air layers.

The ficus Alii has been popular in the U. S. trade for only the last 10 years or so. It is almost always produced as a standard tree, in pots and tubs ranging from 10 to 17 inches (25 to 42.5 cm).

VARIETIES

Fiddle-leaf has a Compacta variety, which is a little smaller plant with shorter internodes, somewhat better suited to bush production. *F. elastica* cv. Decora is the most common rubber tree variety. Its leaves are broader and more shiny than the species, with reddish pigmentation on the undersides of the leaves. Robusta has solid green leaves, somewhat larger than Decora's. It grows a little more aggressively, also, as its name would imply. Burgundy has rich, deep red foliage, which almost turns purple under lower light levels. These three varieties have a reddish sheath covering the emerging leaf. Asahi is a slightly smaller, variegated Decora with green and yellow coloration. Several other green and variegated varieties are also in the trade. In addition to Alii, a new cultivar of *F. maclellandii* called Amstel King has been introduced from Holland. It has somewhat more vigor than Alii, with reddish new foliage.

 The nurseries most likely to go out of business are those which expand too quickly.

PROPAGATION

At one time travelers to South Florida and Central America saw extensive stock plant farms of Decora and Robusta ficus. The extensive plantations generated air layers for foliage farms in the United States and Europe. Today virtually all of that production has shifted to tissue culture, and the cutting farms are now producing other plant varieties. The tissue-culture rubber plants tend to be more compact and juvenile in appearance, with better branching. Also, they don't become overgrown as easily as the vegetatively propagated rubber plants.

F. lyrata is virtually all from tissue culture today, as well, generally for the same reasons. Standard and large trees may be better propagated from cuttings, but the tissue-culture plants make a far superior 10-inch (25-cm) bush. Many commercial tissue-culture labs offer these varieties. *F. elastica* cultivars are also propagated to some degree by leaf-and-eye cuttings in Europe.

CULTURE

The *F. elastica* cultivars generally do better in bright light, around 5,000 f.c. (54 klux). The plants are frequently grown in shadehouses with 47 to 63% shade. Anything darker tends to result in weak, stretched plants, possibly with less contrasting color. Rubber plants need a reasonably well-drained mix, and they do best when irrigated regularly. Don't permit them to dry excessively, or lower leaf loss and dull color will make the plants unattractive. One nice feature of *F. elastica* cultivars is that you don't have to do much spraying.

F. lyrata is usually grown under 63% shade, though it can be produced at somewhat higher light levels. Moisture requirements are a little lower than for the rubber plants, as diseases tend to ensue when *F. lyrata* is overwatered. Soil aeration is quite important in this variety, which needs to be sprayed regularly for foliar disease and mites under shadehouse conditions. Alii is sometimes started in full sun to help build stem caliper prior to moving into shaded areas. Plants are typically finished in 63 or sometimes 73% shade. It generally helps to keep Alii under shade at least several months to acclimate it to the ultimate consumer's interior environment.

NUTRITION

The fiddle-leaf fig has average nutrient requirements. Medium rates of 3-1-2 granular or coated, slow-release fertilizers are generally used. *F. lyrata* is rather fond of trace elements, which should be incorporated into the potting soil or at least into the granular fertilizer. Chlorosis problems from lack of manganese or iron are fairly common, resulting in a fairly severe yellowing of newer leaves. Other than that, it doesn't have any special nutritional requirements, though it needs adequate sulfur for good color.

  F. elastica varieties prefer more fertilizer than *F. lyrata,* and they require more potassium. The grower may do better with a 2-1-2 ratio in winter or in wet climates. The large leaves of *F. lyrata* and *F. elastica* make slow-release fer-

TABLE 16 **Leaf analysis rating standards for *Ficus elastica* cv. Decora**

Nutrient (%)	Very low	Low	Medium	High	Very high
Nitrogen	<1.00	1.00–1.29	1.30–2.25	2.26–3.50	>3.51
Sulfur	<0.10	0.10–0.14	0.15–0.50	0.51–1.00	>1.00
Phosphorus	<0.08	0.08–0.09	0.10–0.50	0.51–1.00	>1.00
Potassium	<0.45	0.45–0.59	0.60–2.10	2.11–3.50	>3.50
Magnesium	<0.15	0.15–0.19	0.20–0.50	0.51–1.00	>1.00
Calcium	<0.20	0.20–0.29	0.30–1.20	1.21–2.50	>2.50
Sodium			<0.21	0.21–0.50	>0.51
(ppm)					
Iron	<20	20–29	30–200	201–1000	>1000
Aluminum			<250	251–2000	>2000
Manganese	<15	15–19	20–200	201–1000	>1000
Boron	<15	15–19	20–50	51–75	>75
Copper	<6	6–7	8–100	101–500	>500
Zinc	<11	11–14	15–200	201–1000	>1000

Sources: Institute of Food and Agricultural Sciences, Apopka, Florida; Dr. Benjamin Wolf, Fort Lauderdale, Florida.
Notes: Common name: rubber tree. Sample of most recent fully mature leaves, no petioles.

tilizer incorporated into the media an attractive option, as top-dressing can be problematic once plants mature.

Nutrient requirements of Alii are similar to those of the rubber plants. They are sensitive to too much boron, so be conservative with boron in the feed program. They occasionally have deficiencies of manganese or zinc, giving you small, narrow, crippled leaves.

DISEASES

Most *F. elastica* cultivars are sensitive to *Cercospora*, a fungus causing tiny leaf spots, with the leaf ultimately turning yellow. *Cercospora* prefers temperatures around 77F (25C), as well as warm days with cool nights. Sprays with thiophanate methyl fungicides or Dithane (mancozeb) control it, but you need to get good coverage on the underside of the foliage.

F. lyrata is affected by several leaf spot diseases. *Alternaria, Phyllostica, Myrothecium, Xanthomonas,* and *Pseudomonas* can all cause spots, most of which are red and very difficult to distinguish. You should have a disease analysis performed if spotting is occurring on *F. lyrata,* in order to know what disease you are fighting. *Pseudomonas* can also cause the midrib and the upper growing point to turn reddish brown, resulting in stem and leaf dieback. When this occurs, trim back to healthy wood with sterilized clippers and spray a copper fungicide. *Fusarium* also occasionally causes a root and stem rot disease in *F. lyrata*. Sprays or a drench with a thiophanate methyl fungicide can be beneficial.

Ficus Alii has relatively few foliar disease problems.

INSECT AND MITE PESTS

Rubber plants suffer from thrips injury where the new leaves are attacked. Milky sap leaks from the lesions, causing staining. Sprays of Orthene (acephate) or Mavrik (tau-Fluvalinate) are helpful, especially with a wetting agent. Mealybugs sometimes attack rubber plants, and controls are similar to those listed in other foliage varieties.

F. lyrata frequently suffers from spider mite activity. Because of the thickness of the leaves, however, two-spotted mites are generally found on only the very immature leaves. They have trouble attacking the thick, leathery, mature leaves. Even one or two spider mites feeding on the tiny, young leaves will cause red spots of varying sizes to develop as the leaf grows out. Sprays of Pentac (dienochlor), Avid (abamectin), or Talstar (bifenthrin) should help control them. Alii is attacked in hot, dry weather by microscopic mites. The mites cause small, crippled leaves with marginal necrosis. Sprays of Avid (abamectin), Kelthane (dicofol), or Thiodan (endosulfan) should control the microscopic mites. Predator mites *(Phytoseiulus)* are useful on *F. lyrata* as a biological control.

DISORDERS

Opaque, somewhat rectangular spotting can develop on the new leaves of Decora cultivars. This symptom has been attributed to moisture stress and high light conditions, especially when air-layering. Rubber plants grown close together in shadehouses may suffer from severe leaf drop in heavy rains, due to ethylene generation from waterlogged soils. Spacing helps dissipate the ethylene.

Decora leaf stripe can be caused by a foliar nematode, *Aphelenchoides*. The nematode has an interesting way of getting to the foliage of the large plants. The nematode attaches itself to smutgrass or similar weeds. As the weeds grow up, they come into contact with the rubber plant foliage, upon which the nematode is parasitic. Keeping weed populations down largely eliminates the problem.

Crippled, tiny new leaves in rubber plants are usually caused by lack of boron. Foliar boron sprays return subsequent growth to normal. When exposed to cool temperatures or chilling winds, the undersides of Decora leaves will become a blush red color. Sprays of potassium nitrate help the leaf color return to normal. When the growing tips of rubber plants are very short, it suggests either lack of fertility or root disease problems.

 Unfertilized spider mite eggs hatch into male mites.

F. lyrata, as indicated earlier, develops red spots because of numerous possible causes. The shape and distribution of the red spots are helpful in diagnosis, but lab confirmation is usually necessary. Lack of moisture contributes to leaf drop, as does ethylene from waterlogged soils or heaters which do not completely combust the propane. Alii suffers leaf drop and twig dieback if allowed to dry out. Keep a steady supply of moisture to it, as you would for *F. benjamina*.

TRICKS

It is very difficult to sell *F. lyrata* and Decora cultivars once they become over-grown. Applying growth regulators before that happens will significantly increase the time frame in which the plants can be sold. Foliar sprays of B-nine (daminozide) at 2,500 ppm are very effective in controlling their height. B-nine also helps improve color. Sprays of potassium nitrate help reduce red-dening of *F. elastica* cultivars caused by exposure to winter temperatures. The sap of *F. lyrata* oxidizes when exposed to the air. Therefore, anything that causes injury to the leaves, including insect feeding, foliar disease, and spray injury, will result in a red spot.

The grower should regularly inspect the emerging *F. lyrata* leaves (which are about the size of your thumbnail) for mite or disease activity. Use a wet-ting agent or surfactant when spraying for mites on *F. lyrata* and Alii, and make sure you get adequate coverage on the emerging leaves.

Pinch out the growing tips of Alii early in order to keep the plant extra bushy. Pinching, trimming, and keeping Alii pot bound all help encourage the development of good, thick trunks.

INTERIOR CARE

The *F. elastica* cultivars have relatively few indoor problems. They do better in relatively high light, 200 f.c. (2.2 klux) and higher. They tolerate 75 to 100 f.c. (0.8 to 1.1 klux), but the leaves tend to droop. Rubber plants may stretch badly in interior environments if they are fertilized. Don't keep them too wet, but don't allow them to dry out, either, or older leaves will begin to turn canary yellow and then brown. Rubber plants become root bound fairly easily in interior environments. Give them little if any fertilizer, or they may become too large. Rubber plants can be pruned and shaped, when necessary.

F. lyrata performs fairly well indoors. Watch especially for spider mite activity, once again concentrating on the emerging foliage. *F. lyrata* requires less water than *F. elastica* cultivars, as it is much less succulent. Light require-ments are similar to those of rubber plants. Avoid excessive drying on Alii indoors, as leaf drop and twig dieback can be a problem. Prune out dead twigs carefully if this occurs.

REFERENCES

Broschat, T.K., and H.M. Donselman. 1981. Effects of Light Intensity, Air-Layering and Water Stress on Leaf Diffusive Resistance and Incidence of Leaf Spotting on *Ficus elastica*. *Hortscience* 16 (2): 211-212.

Chase, A.R. 1988. Fertilizing Ficus. *Florida Nurseryman* (July): 57-59.

Conover, C.A., and R.T. Poole. 1978. Production of *Ficus elastica* 'Decora' Standards. *Hortscience* 13 (6): 707-708.

Conover, C.A., and R.T. Poole. 1983. Shade and Fertilizer Affect Fiddle-Leaf Fig. *American Nurseryman* (July 15): 35-36.

Marlatt, R.B. 1979. The Leaf Nematode Disease of *Ficus elastica* 'Decora'. *Foliage Digest* (August): 7.

Osborne, L.A., R.W. Henley, and A.R. Chase. 1983. Foliage Plant Research on *Ficus*. *Nursery Notes* (August): 2-6.

Poole, R.T., and C.A. Conover. 1990. Propagation of *Ficus elastica* and *Ficus lyrata* by Cuttings. *Foliage Digest* (July): 3-4.

Powell, C.C., and R. Rossetti. 1992. *The Healthy Indoor Plant*, 214-215. Athens, Ga.: Rosewell Publishing, Inc.

FIDDLE-LEAF FIG

(*See* Ficus lyrata)

FISHTAIL PALM

(*See Other Foliage:* Caryota)

GARDENIA

HABITAT

Gardenias may or may not be considered foliage plants, as they are primarily produced for cut flowers and landscaping. However, their foliage is attractive enough that they would be grown commercially even if they didn't produce such brilliant, sweet-smelling flowers.

Most of the 200 species of gardenias grow as small shrubs or trees, and most are native to Old World tropics and subtropics. The most commonly cultivated gardenias are varieties of *Gardenia jasminoides*, which is native to Zone 8 in China. It was once thought that gardenias came from the Cape of Good Hope, hence the common name "cape jasmine."

Fig. 26 Gardenia from the author's collection. (Photo by the author)

Gardenia thunbergia, which is commonly used as root stock for grafted landscape gardenias, is actually native to South Africa. The genus name *Gardenia* comes from A. Garden, a Charleston, South Carolina, doctor, and *jasminoides* refers to the jasminelike flowers.

USES

Gardenias have deep green, somewhat shiny foliage, with a compact habit and dense growth. They generally bloom naturally in spring and summer, typically from March through June, but occasionally later, depending on latitude. Foliage growers produce gardenias in containers ranging from 5 to 10 inches (12.5 to 25 cm). One cutting per pot is generally used, and the plants are trimmed for fullness. Their primary sales periods are Christmas, Easter, and Mother's Day, with some sales potential throughout the rest of the year.

Gardenia radicans is occasionally grown as a hanging basket. Different varieties are also produced as shrubs and ground covers for landscape use, especially in the South, and for cut flowers.

VARIETIES

Mystery, probably the most familiar gardenia variety, has large flowers approximately 4 inches (11 cm) across. The plant can grow to 6 to 8 feet (1.8 to 2.4 m) as a mature landscape specimen. Miami Supreme and Glazeri, two other popular cultivars, have characteristics somewhat similar to Mystery. August Beauty also has large flowers, though it is generally a smaller plant. Veitchii is a smaller, early-blooming variety with medium-sized flowers. *G. radicans* is one of the miniature gardenias, with a low, spreading habit and small- to medium-sized blooms. All of these varieties have deep green foliage with brilliant white, fragrant flowers.

PROPAGATION

Almost all gardenia propagation is from cuttings. They can be taken anytime, though spring and summer propagation is preferred in Florida; in northern areas December to January propagation is popular. Tip or midsection cuttings of crisp, half-mature stem tissue, six to eight weeks old, are taken. The cuttings are usually 4 to 5 inches (10 to 12.5 cm) long, with two to three sets of leaves. The ideal propagation temperature is 70F (21C). It helps to propagate with some shade in order to protect the cuttings from moisture stress.

Fifty-fifty mixtures of peat and sand or peat and perlite are popular propagation media. Bottom heat helps speed rooting in cool situations. Regular, intermittent misting, typically a few seconds every 10 minutes or so, is used. Rooting generally takes six to eight weeks.

Tent propagation is also practiced. Gardenias grown for landscaping, furthermore, may be grafted onto *Gardenia thunbergia,* which is somewhat nematode resistant.

CULTURE

As they are not tropical plants, gardenias prefer cooler temperatures, especially at night. Cool night temperatures are important for bud development. Night temperatures of 61 to 65F (16 to 18C) are desirable, though some growers prefer to maintain 50 to 55F (10 to 12.8C) at night. Optimum day temperature is about 70F (21C), but healthy gardenias will handle much higher temperatures when growing vegetatively.

The plants need to be cut back at least one or two times in order to have adequate fullness for good bloom count. The growing medium is usually a mixture of peat and bark with a little bit of sand. Gardenias are acid loving plants, so the potting medium can be adjusted to a pH of 5 to 5.5. For symmetry and even flowering, gardenias like air movement and need to be well spaced to get good light all around. Spacing is approximately 18 inches (45 cm) on center, depending on container size and variety.

Providing gardenias with four weeks of short days can help with flowering. Good light levels in winter are important.

NUTRITION

Acid-loving plants are really iron-loving plants. Gardenias and other so-called acid-loving plants can grow perfectly well at higher pH values if they receive adequate available iron. Acidic, granular fertilizer formulations with a 1-1-1 ratio of $N-P_2O_5-K_2O$ are popular for gardenias, with some of the nitrogen coming from ammonium sulfate or sulfur-coated urea. Soluble 1-1-1 ratio fertilizers, such as 20-20-20, are popular, especially for smaller containers, where top-dressing is labor intensive. Constant liquid feed of 150 to 200 ppm nitrogen is also used, and it helps to rinse the liquid fertilizer off the foliage. An occasional leach, every 10 days or so, is desirable to avoid buildup of soluble salts. Trace elements, with emphasis on iron, should be incorporated into the potting medium.

Iron chlorosis is quite common in gardenias, resulting in significant yellowing of the plant. It happens especially in alkaline or poorly aerated soils or when root disease or nematode problems exist. Powdered sulfur can be drenched at 3 pounds per 100 gallons (1.4 kg per 400 l), or you can use iron sulfate at 1 to 2 pounds per 100 gallons (0.49 to 0.9 kg per 400 l). Iron chelate drenches and sprays are also beneficial. Plants lacking in nitrogen will have overall pale green color and little growth. Lack of phosphorus results in

TABLE 17 Leaf analysis rating standards for *Gardenia jasminoides*

Nutrient (%)	Very low	Low	Medium	High	Very high
Nitrogen	<1.2	1.2–1.4	1.50–3.00	3.01–4.00	>4.00
Sulfur	<0.13	0.13–0.19	0.20–0.40	0.41–0.75	>0.75
Phosphorus	<0.12	0.12–0.15	0.16–0.40	0.41–0.75	>0.75
Potassium	<0.80	0.80–0.99	1.00–3.00	0.31–4.50	>4.50
Magnesium	<0.20	0.20–0.24	0.25–1.00	1.01–1.25	>1.25
Calcium	<0.31	0.31–0.49	0.50–1.30	1.31–2.50	>2.50
Sodium			<0.21	0.21–0.50	>0.50
(ppm)					
Iron	<40	40–49	50–250	251–2000	>2000
Aluminum			<251	251–1000	>1000
Manganese	<40	40–49	50–250	251–1000	>1000
Boron	<20	20–24	25–70	71–100	>100
Copper	<4	4–5	6–40	41–500	>500
Zinc	<16	16–19	200–150	151–1000	>1000

Sources: Institute of Food and Agricultural Sciences, Apopka, Florida; Dr. Benjamin Wolf, Fort Lauderdale, Florida.

Notes: Common name: gardenia. Sample of most recent fully mature leaves, no petioles.

smaller-than-usual, dark green leaves. Plants low in potassium have burned leaf margins, especially in the older leaves. Potassium is also important for flowering.

DISEASES

The most troublesome gardenia disease is canker, caused by the fungus *Phomopsis gardeniae*. The disease begins as slightly sunken, discolored areas on the lower stem and perhaps below soil level. The disease progresses to form rough cracks on the stems near the soil line, and cankers develop. On these areas numerous spores of the fungus form and spread by splashing water and handling of plants. The disease tends to enter through wounded stem tissue, so avoid wounding at all costs. Fungicides are not effective as a cure of this disease, and infected plants should be discarded. Sprays of copper fungicides can help prevent spread of canker.

Oxygen is essential for iron availability.

Several leaf spots, including *Anthracnose, Alternaria, Cercospora,* and several others, can affect gardenias. Copper fungicides are generally used against these pathogens. *Botrytis* can attack foliage and buds, especially spent flowers. Copper fungicides are generally used here, though some of the carbamate fungicides are also registered. Powdery mildew is an occasional problem, and sprays of Bayleton (triadimefon) are helpful. Gardenias are also rather sensitive to nematodes, especially in the landscape.

INSECT AND MITE PESTS

Unfortunately, numerous insect pests attack gardenias. Thrips are frequently seen in the flowers. Sprays of Mavrik (tau-Fluvalinate) with a wetting agent or surfactant can help, especially when applied while the buds are forming. Three types of aphids commonly attack gardenias: the melon aphid, the foxglove aphid, and the lily aphid. Numerous insecticides are registered for aphids, including Diazinon, Cygon (dimethoate), insecticidal soaps, and many others.

Several scale insects also cause problems, which include twig dieback, reduced leaf size, and lack of vigor and flowering. Sprays of Cygon (dimethoate), Dycarb (bendiocarb), and horticultural oils can be helpful, though you usually need to make two or three applications. Granular Di-syston (disulfoton) is also very good. Mealybugs are frequently found at the bases of the leaves. They cover themselves with their eggs and cottony wax. Chemical controls are similar to those for scale insects; again, repeated applications are usually necessary. Both red and two-spotted mites attack gardenias, causing speckling and discoloration, as well as reduced vigor. Miticide choices include insecticidal soaps, horticultural oils, Pentac (dienochlor), and Avid (abamectin).

Whiteflies can also be particularly troublesome, especially with the spread of new strains of whitefly in recent years. Sprays of horticultural oils, insecticidal soaps, mixtures of Orthene (acephate) plus Talstar (bifenthrin), and various neem-extract insecticides are helpful. Granular Marathon (imidacloprid) is also very good against whitefly, as well as aphids, mealybugs, and thrips. Finally, the fuller rose beetle eats notches in the leaves. The larvae feed on the roots, which can ultimately kill the plant. Sprays and drenches of Orthene (acephate) may help, but this insect is difficult to control.

DISORDERS

A primary disorder is bud drop. Flower loss has many causes, most of which are physiological, including root injury, nematode activity, excessive soil moisture, high fertility rates, and insect activity. Hot, dry weather also contributes to bud drop, as does a rapid drop in temperature. Growers have to examine a bud drop situation carefully to determine what is causing the bud drop. Any handling or even touching of the petals causes the flowers to turn brown.

Injury from Cycocel (chlormequat) causes irregular yellowing or bleaching of the foliage. It generally goes away in a few weeks. Windy conditions can very quickly turn the older leaves of gardenias a canary yellow. Black, sooty mold can form when insects are feeding heavily on plant sap. Applying insecticides is necessary, and horticultural oil sprays help to loosen and kill the sooty mold. Finally, large, tan lesions can develop from sunscald when strong sunlight shines on wet foliage.

TRICKS

Gardenias don't care for wind, but they do like fresh air and some air movement. They seem to struggle under stuffy conditions. They respond well to carbon dioxide enrichment in closed greenhouses, usually at the rate of 1,000 ppm. Sprays of B-nine (daminozide) are very useful in increasing color and keeping the plants compact. The more compact growth tends to result in higher bloom density, though flowering may be delayed slightly. B-nine also tends to stop bypass growth, where vegetative growth after bud set obscures the flowers. Spray B-nine at 5,000 ppm when plants are two-thirds grown, from mid-August to mid-September in Florida, or the first two weeks of December for other areas.

Gardenias should be pruned to have 6- to 8-inch (15- to 20-cm) shoots by October 1 for spring-flowering crops. Pruning after October 1 tends to reduce the number of blooms the following spring. Intermittent lighting can speed bud opening, as can artificial night lighting. Lighting can be applied for a period of four to six hours at night.

Keep night temperatures low to permit buds to form. Buds can initiate under many temperature and light regimes, but cool temperatures at night are required for them to develop. Do not spray mancozeb fungicides, such as Dithane, on gardenias, or severe phytotoxicity may result.

INTERIOR CARE

Flowering gardenias placed in interior situations usually maintain their blooms for only a few weeks or so. They will generally not bloom inside after that. In more northern climates gardenias are frequently kept outside for flowering in the summer, then brought inside for winter protection.

They should not be fertilized indoors, or severe stretching will occur. If you do fertilize, use one of the acidic fertilizer blends to help maintain color. Gardenias are not fond of alkaline irrigation water, so collecting rainwater may be preferred. Sulfur can be applied carefully as a drench or topdress to help lower soil pH, also. Use about one-third of a teaspoon per gallon (3.8 l) when drenching with sulfur. Be careful, though, as this can increase soluble salts.

Gardenias prefer cool, bright conditions indoors. Keep gardenias in as high a light level as you can when maintaining them indoors, and watch out for insect pests.

REFERENCES

Arthur, J.M., and E.K. Harvill. 1937. Forcing Flower Buds in *Gardenia* With Low Temperature and Light. *Boyce Thompson Institute for Plant Research* 8 (5): 405-412.

Black, R.J. 1980. Gardenia and Hibiscus Flower Bud Drop. *The Florida Nurseryman* (September): 113.

Clark, B. 1984. Gardenias Bring Questions About Problems, Culture. *Southern Florist and Nurseryman* (May): 91.

Conover, C.A. 1980. *Gardenias in Florida*. IFAS Circular 313 (September): 1-8. Gainesville: University of Florida Agricultural Extension Service Institute of Food and Agricultural Sciences.

Conover, C.A., T.J. Sheehan, and R.T. Poole. 1968. Flowering of Gardenia as Affected by Photoperiod, Cycocel, and B-nine. *Florida Flower Grower* 5 (12): 1-5.

Watkins, J.V., and T.J. Sheehan. 1975. *Florida Landscape Plants*, 392. Gainesville: The University Presses of Florida.

GLOXINIA

(See Other Foliage: Sinningia*)*

GRAPE IVY

(See Cissus*)*

HEDERA

English Ivy

HABITAT

Hedera was the classical name for ivy. The plant is commonly called English ivy because it was brought to America by English settlers. Ivies have been used in landscapes and as houseplants for many centuries. Most are trailing vines, typically with leaves that have three to five lobes. Juvenile foliage is different from mature foliage. The adult leaves are more rounded and elongated and are generally seen only in mature landscape plantings.

Juvenile shoots mutate rather freely, so there are hundreds of cultivated varieties. Curiously, though, there are only five species of ivy, which is in the Aralia family. Most cultivated varieties are selections of *Hedera helix*. The Algerian ivy, *Hedera canariensis,* is occasionally produced commercially. Ivies are native to Southern Europe and Northern Africa, primarily. Being from a Mediterranean climate, they prefer mild temperatures and tolerate low humidity fairly well. When well rooted they are fairly drought tolerant, and most are very cold hardy.

USES

Ivies are popular as small potted plants for desks, coffee tables, plant stands, and windowsills. Smaller pot sizes typically range from 3 to 6 inches (7.5 to 15 cm). Because of their trailing habit,

Fig. 27 English ivy in a window box. (Photo by the author)

ivies are widely produced as hanging baskets, with 6- and 8-inch (30-cm) containers being the most popular sizes. The plants can also be grown in dish gardens, on trellises, or as topiary. Hederas are used for ground covers in interior and exterior landscaping and in combinations as basal plants for large trees.

VARIETIES

Most of the new ivy cultivars are self-branching, which helps them fill in a pot or hanging basket quickly. Green varieties include the species type as well as Asterick and the California series. Some white-variegated cultivars include Glacier, Ingelise, and Kolibri. Yellow-variegated types include Gold Dust and Gold Child, which are nice plants, though rather soluble salt sensitive.

Ivy cultivars vary in their susceptibility to spider mites, bacterial leaf spot, and soluble salts. The variety California, for example, has high susceptibility to spider mites but low susceptibility to *Xanthomonas*. Some good cultivars for quick finishing include Manda's Crested, California Weber, and Curlilocks.

PROPAGATION

English ivies are generally very easy to propagate by cuttings. Juvenile leaf-and-eye cuttings can be rooted all year. In fact, in 1962 Charles Hess isolated compounds called rooting cofactors from the juvenile wood of English ivy. A blend of indoleacetic acid and rooting cofactors was found to have a synergistic effect on the rooting of cuttings of other species. Rooting hormones are generally not required because of the ease in rooting. During warm periods the grower may like to mist fresh ivy cuttings every half hour or so.

Seed propagation is occasionally done on ivy, though not usually on a commercial basis. Flowers can occur on mature wood in September and October. The blackish fruits are frequently poisonous.

 Boron is the only trace element which leaches significantly.

Algerian ivy, *Hedera canariensis,* can be rooted any time of the year, once the new flush of growth has become firm. The stems should be at least partly crisp. Frequently, 1,000 ppm IBA (indolebutyric acid) is used. Virtually all of the cuttings should root under mist within four or five weeks.

CULTURE

Hederas are generally produced in greenhouses with 1,500 to 2,500 f.c. (16 to 27 klux) of light. In warm climates 1,800 f.c. (19.4 klux) is a popular level, though the variegated varieties usually need more light. Under low light situations ivies may lose variegation, and they tend to become leggy.

Optimum temperatures range from 65 to 80F (18.3 to 26.7C). Northern growers may find it more economical to produce ivy between 58 and 65F (14.4 and 18.3C). Try not to let the temperature get above 90F (32.2C). Hederas are generally very cold hardy, with most commercial cultivars tolerating down to 10 to 20F (-12.2 to -6.7C). Some ivies can even handle -10F (-23.3C).

The medium should be well drained, have a reasonably high moisture-holding capacity, and have a pH of about 6.0. Many growers use six or seven plants per pot for an 8-inch (20-cm) container, four or five plants for a 6-inch (15-cm) container. You can use fewer plants per pot, then pinch for fullness, though production time will be a little longer.

NUTRITION

It is best to fertilize ivies conservatively because they are more sensitive to soluble salt injury than many other foliage varieties. Six grams of slow-release 19-6-12 or 18-6-8 per 5-inch (12.5-cm) pot every three months will do a good job. You may want to use less on some of the more sensitive cultivars. Constant liquid feed of about 150 ppm nitrogen from 20-10-20 or 9-3-6 is frequently used. It helps to rinse after liquid feeding to avoid foliar burn. Fertilization should be especially gentle in winter.

Research has shown that the nitrogen source is not particularly important for ivy, though northern growers tend to use more nitrate sources, especially in winter. Ivies are not particularly prone to trace element deficiencies. It is usually not necessary to incorporate minor elements into the medium; a standard trace element package in the fertilizer program is usually sufficient. Variegation contrast may be increased by maintaining good levels of magnesium.

DISEASES

The most troublesome disease of ivy is bacterial leaf spot, which has been known for at least 60 years. *Xanthomonas* causes brown to black, small leaf spots, most commonly on older foliage. The spots have no visible fruiting bodies, though they tend to have a yellow halo or a water-soaked edge.

Sanitation and good culture are the best controls, and it certainly helps to keep the foliage dry. Sprays of copper fungicides or Aliette (fosetyl-aluminum) help, but don't spray them close together in the rotation, or phytotoxicity can result.

Several fungal diseases attack ivy, including anthracnose, caused by the fungus *Colletotrichum*. Large, roundish, black leaf spots form, frequently with black, visible fruiting bodies on the underside of the leaf. In addition to cultural controls, sprays with Dithane (mancozeb), Chipco 26019 (iprodione), or one of the thiophanate methyl fungicides control it.

Phytophthora causes the inner leaves to turn brown and curl downward. Large gray to black leaf spots may also form, especially where the petiole joins the base of the leaf. Dry the plants out if this occurs, then apply a soil drench of Truban (etridiazole) or Subdue (metalaxyl). *Pythium* symptoms and treatments are similar to those of *Phytophthora*.

Rhizoctonia causes a very sudden, severe blight, with many brown leaf spots, especially in summer. Sprays of Ornalin (vinclozolin), Chipco 26019, or a thiophanate methyl fungicide, such as Cleary's 3336 or Domain, help. In cool weather *Botrytis* causes large, gray lesions. Clean up any old leaves and debris, watch your irrigation carefully, and spray with either Ornalin or Chipco 26019.

INSECT AND MITE PESTS

The two-spotted mite is the major pest of ivy. It causes speckled foliage and leaf yellowing, and it may create marginal burn symptoms in variegated types. Predator mites *(Phytoseiulus)* are useful, as are sprays of Mavrik (tau-Fluvalinate) or Avid (abamectin). Broad mites cause distorted new growth, and the growing tip may die back completely. Remember that broad mites are not visible to the naked eye. Sprays of Kelthane (dicofol) or Pentac (dienoclor) are useful, especially when combined with a surfactant.

Caterpillars occasionally chew holes in leaves, especially during the spring months. Leaf rollers also cause chewing injury. Both can be controlled with sprays of Dipel *(Bacillus thuringiensis)*, Orthene (acephate), or Dursban (chlorpyrifos). Mealybugs tend to concentrate in the leaf axils, where their feeding creates honeydew, upon which sooty mold may form. Systemic insecticides are usually best here, including Cygon (dimethoate) or granular Di-syston (disulfoton). Insecticidal soaps also work fairly well, especially with repeated applications. Root mealybugs are occasionally found in the soil, feeding on roots and reducing vigor. A soil drench of Diazinon should take care of them.

Disorders

Soluble salt injury causes severe burn, usually on most or all of the leaves. The burn tends to be worse on the more highly variegated varieties. The *Hedera* grower should fertilize conservatively and monitor soluble salts on a regular basis. Spraying with Agrimycin 17 (streptomycin sulfate) tends to cause pale yellowing and whitening in new leaves. Most varieties are susceptible, so it is safest not to use this product for bacterial leaf spot control on ivy. *Hedera* is sensitive to bromine from Agribrom at high rates; 55 ppm bromine can injure ivy. When using Agribrom for algae control, keep it at 25 ppm on ivy. Spraying with Daconil (chlorothalonil) can cause necrosis and distortion of new leaves of hederas, so it is best to avoid it on this plant.

Many ivy cultivars are somewhat unstable genetically. Problems with reversion can occur on container production or in stock beds. Variegated types may revert to all green, or the variegation pattern or even the leaf shape may change. The best treatment is to immediately prune out any undesirable or reverted growth. Variegation is best maintained by good light, adequate magnesium, and conservative nitrogen fertility.

Tricks

When liquid-feeding ivy make sure the entire soil mass is moistened to avoid localized buildup of soluble salts. After an occasional leach to keep salts down, try to dry ivy out thoroughly before resuming irrigation. Spider mites and their injury symptoms may be difficult to see in variegated varieties. Growers should inspect hederas for spider mites regularly. Some like to spray preventatively with a miticide every couple of weeks or so, especially in warm or dry weather. Ivy production is also a good place to try a predator mite program.

Make sure stock plants get adequate magnesium, or variegation will be reduced with time. Occasional feeding with calcium nitrate helps keep high-quality vines for rooting. Sprays or dips with a 1% solution of household vinegar aid in controlling bacterial leaf spot in container plants, stock beds, and cuttings.

Interior care

Ivy prefers bright, indirect light indoors. A minimum of 150 to 250 f.c. (1.6 to 2.7 klux) is preferred. A few shade-tolerant types can be maintained at lower light levels. The variegated varieties generally do better with more light. Temperatures should be maintained between 60 and 75F (16 and 24C). Some ivies tolerate down to 35F (2C) or lower, but for good, continuous growth, it is safest to not let the temperature drop below 45F (7C).

The older leaves of ivies tend to turn yellow and then brown if the plants are allowed to become too dry. It helps to dry ivy down occasionally, but don't let ivy go too long, or leaves will be lost. Occasional gentle fertilization, usually from soluble 20-20-20, helps maintain leaf size. Inspect plants occasionally for spider mite activity, especially when new plants are placed in interiors.

REFERENCES

Blessington, T.M., and P.C. Collins. 1993. *Foliage Plants: Prolonging Quality*, 131-134. Batavia, Ill.: Ball Publishing.

Chase, A.R. 1989. Nitrogen Source and Rate Affect Severity of Xanthomonas Leaf Spot of *Hedera helix. Nursery Digest* (April): 18-19.

Chase, A.R. 1994. Common Diseases and Disorders of English Ivy. *Landscape and Nursery Digest* (October): 64-65.

Chase, A.R., and J.M.F. Yuen. 1986. *Effect of Fertilizer Rate on Growth of Six Cultivars of* Hedera helix. AREC-Apopka Research Report RH-86-20: 1-3. Apopka: Agricultural Research and Education Center.

Dirr, M.A., and C.W. Heuser. 1987. *The Reference Manual of Woody Plant Propagation*, 131. Athens, Ga.: Varsity Press Inc.

MacCubbin, T. 1991. Ivies Gain Popularity. *Florida Foliage* (August): 26-27.

Osborn, L.S., and A.R. Chase. 1985. Susceptibility of Cultivars of English Ivy to Two-Spotted Spider Mite and Xanthomonas Leaf Spot. *Hortscience* 20 (2): 269-271.

Osborn, L.S., A.R. Chase, and R.W. Henley. 1984. English Ivy. *Nurserymen's Digest* (December): 71-76.

HELICONIA

HABITAT

Perhaps no other plants symbolize the tropics better than heliconias. They look like rambling banana plants, their lush, succulent foliage and colorful, exotic flowers giving them the ultimate tropical look. They were formerly listed in the banana family, but they now have their own taxonomic family, Heliconiaceae. The number of species approaches 300, and new selections are being discovered and bred constantly. Com-mon names include "lobster claw" and "false bird of paradise."

The plants spread by a fleshy, underground rhizome. After they grow four to seven leaves, two or more boat-shaped bracts emerge from a central floral axis. The subsequent flowers are exotically shaped and waxy in appearance, ranging from yellow, orange, red, pink, and white, to black, with many varieties having multiple colors on one bloom.

Many heliconias are native to northern South America, including *Heliconia psittacorum*. Many are found, for example, growing at sea level in French Guiana in 12,000 to 14,000 f.c. (129 to 151 klux) of equatorial light. They prefer naturally high light levels and plenty of moisture and have almost no cold tolerance. In the equatorial habitat heliconias are not photoperiod sensitive, so they have the capacity to flower all year long.

Fig. 28 *Heliconia* Golden Torch in constant bloom. (Photo by the author)

182

USES

Heliconias have been in the U.S. foliage trade since about 1984, longer in Europe. Smaller cultivars, especially those of *H. psittacorum*, can be grown in 6- and 8-inch (15- and 20-cm) pots. Many of the larger varieties are grown in 10-inch (25-cm), 14-inch (35-cm), and even larger containers. Container size largely depends on the size of the cultivar and the market.

Heliconias create a striking tropical effect when installed in mass plantings in well-lit interiors. They make ideal specimens for atriums, attractive when placed in large urns. The smaller varieties perform almost like a flowering ground cover, whereas large cultivars create more of a jungle effect. Several *Heliconia* varieties are also produced as cut flowers.

VARIETIES

The most familiar small heliconia cultivars are strains of *Heliconia psittacorum*. Andromeda was found at Andromeda Gardens in Barbados in 1978 by my friend Dr. Al Will. Its origin is unknown, but the red-and-orange flowers are quite well known today. Dr. Will also found the familiar Golden Torch, which is a cross between *H. psittacorum* and *H. latispatha*. It is a common smaller cultivar with waxy, upright, golden flowers. Choconiana is a good bloomer with yellow-orange, upright flowers. Lady Di is slightly larger, with red bracts and black-tipped, yellow flowers.

One of the most striking large varieties is *H. rostrata*, the lobster-claw heliconia. It easily reaches 6 to 8 feet (1.8 to 2.4 m) and sports foot-long, waxy, red-and-white flowers. *H. caribea* is also very striking; it is a large, upright plant with drooping, red-and-yellow blooms. Several varieties of *H. stricta* and *H. latispatha* are also in the trade. *H. angusta* Holiday once showed promise as a flowering plant for Christmas and Valentine's Day. However, its extreme susceptibility to disease has virtually eliminated it from the trade.

PROPAGATION

Rhizome division and tissue culture are the most common commercial propagation methods. Rhizomes with several eyes can be collected and planted in pots or stock beds, usually 3 to 4 inches (7.5 to 10 cm) deep. Growers frequently place four or five rhizomes or plantlets in a 10-inch (25-cm) container, three or four in an 8-inch (20-cm) container, or three in a 6-inch (15-cm) pot. Occasional misting or fog is helpful in reducing moisture stress until the rhizomes begin to root and grow. Propagation is frequently done under 40 to 60% shade, with optimum temperatures of 80F (27C) during the day and 70F (21C) at night. Once the rhizomes have put out a couple of leaves, the light levels should be increased for best flowering.

Tissue-culture plantlets are popular for production in smaller pots. The number of plantlets will depend partly on the cultivar, but in general it is best to plant heliconias fairly full. Heliconia flowers are frequently pollinated by hummingbirds in the wild, so unless you happen to have hummingbirds in your greenhouse, seed will not be produced.

CULTURE

To support their aggressive growth, heliconias require huge amounts of fertilizer. The more light and fertilizer you give them, the more they will flower. Full sun is best for most cultivars, though some growers like to start the plants in full sun, grow them to the four- or five-leaf stage, then move them under shade. Lower light increases leaf color but reduces flowering.

Heliconias can be started pot to pot, then spaced as they begin to grow. Ideal spacing is 30 to 36 inches (75 to 90 cm) for 10-inch (25-cm) containers, 24 inches (60 cm) for 8-inch (20-cm) pots, and 12 inches (30 cm) of spacing for 6-inch (25-cm) pots. It is best to use full-sized containers rather than azalea pots. The plants appreciate the extra soil, and the container will be more in proportion to the finished plant.

Optimum temperatures are 70 to 90F (21 to 32C). Growth begins to slow below that, and cold damage occurs at 50F (10C). Plant appearance is much better if they are protected from wind. Heliconias need at least a moderate amount of soil aeration. A workable mix might be a 5-1-4 ratio by volume of bark, sand, and sedge peat. Ideal pH range is 4.5 to 6.5. These plants require water frequently and need a potting mix that can handle that. Heliconias have determinate flowers and should produce a bloom after four to seven leaves emerge. Production time ranges between 12 and 24 weeks, depending on container size and variety.

NUTRITION

Nutrient requirements are about the highest I have ever seen for any plant. Constant liquid feed of 300 ppm via overhead or drip irrigation is convenient. I prefer a 2-1-2 ratio of N-P_2O_5-K_2O, such as 20-10-20. Slow-release, coated fertilizers are also very good, especially when incorporated into the media.

Granular fertilizers can be used, but you have to reapply them frequently because of the high watering requirements of the plants. Heliconias are tolerant of elevated soluble salts from high feed rates if produced under high light. Soluble salt tolerance is lower under shade.

It helps to incorporate trace elements into the media prior to planting, and heliconias have a high requirement for magnesium. Be sure you have adequate magnesium in your liquid feed, as most soluble fertilizer preparations do not have enough. If incorporating slow-release fertilizer, you may need to supplement with extra dolomite or EMJEO. Epsom salts will be relatively short lived under frequent irrigation.

Plants quickly become pale when soil nitrogen is inadequate. Iron deficiency may occur as a uniform yellowing in the new leaves. Granular iron applied as a topdress will quickly correct it, as will drenches with chelated iron. Manganese deficiency also causes chlorosis of new leaves, but you also see transverse necrotic streaking or brown stripes in the chlorotic leaves. Sprays of manganese sulfate or chelated manganese generally fix this. Lack of magnesium frequently appears in older plants as broad, yellow margins in the older leaves. Magnesium sprays help some, but since older leaves don't absorb foliar nutrients very well, the magnesium is best applied to the soil.

Potassium often leaches out faster than nitrogen in foliage production.

DISEASES

Helminthosporium is by far the most common leaf spot of *Heliconia*. Round, dry, medium-sized leaf spots form on leaves of any age. It is not a very serious problem, and sprays of Dithane (mancozeb) control it. Usually all you need to do is shift your watering to the middle of the day so that the foliage dries fairly quickly. *Phytophthora* root rot has been reported in Hawaii, but it is fairly rare elsewhere. Systemic *Cylindrocladium* has caused a terrible fungal blight on the Holiday variety. It is characterized by severe leaf browning and dieback, with discolored vascular tissue. Sprays of Terraguard (triflumizole) or Phyton 27 (picro cupric ammonium formate) help control it.

Cucumber mosaic virus has been found on Andromeda and Golden Torch. Longitudinal, yellow mottling following the veins of the leaves occurs, as does pale yellow, water-soaked leaf spots. The virus may be present in some varieties but not become noticeable until the plants are stressed by cold or lack of drainage. There are no controls available, and it may be spread mechanically by cutting tools, mites, or other means.

Nematodes, especially root knot nematodes, can also be troublesome. At the time of this writing, no chemical nematicides are available for nurseries in most of the United States, but hopefully by the time you read this, that will have changed. Hot water dips of plants or rhizomes are useful. Plant material can be placed in water at 100 to 125F (37.8 to 51.5C) for 15 to 30 minutes. The plants or rhizomes should then be placed in cool water immediately afterward, or plant damage would result.

INSECT AND MITE PESTS

Red spider mites are fairly common pests of heliconias, though they are not terribly serious. It can be difficult to spray for them due to the density of the foliage. Predator mites (*Phytoseiulus)*, insecticidal soaps, or miticides such as Avid (abamectin) should control them. Aphids like to attack the open flowers, feeding on the sweet juices that the blooms exude. Numerous insecticides are effective, including insecticidal soaps, Sevin (carbaryl), and Orthene (acephate).

Mealybugs are only an occasional problem for *Heliconia* growers, but control is difficult because there are many places for a mealybug to hide on a *Heliconia* plant. Granular Di-syston (disulfoton) helps control them, as does Cygon (dimethoate) or Diazinon. Thrips may occasionally feed on unfurling leaves or flowers. Sprays of Mavrik (tau-Fluvalinate) or Orthene (acephate) usually work if thrips are a problem.

DISORDERS

Heliconia leaves roll up longitudinally when the plants are dry or moisture stressed. If soil moisture is good, excessive leaf curl may indicate root disease or nematode problems. If the leaves quickly return to normal after irrigation, the roots are probably healthy.

Cold damage first appears as small, black spots on the foliage and the flower stalk. More severe cold causes blackening of aboveground leaf tissue. If temperatures hit freezing, the plants will die back to the ground. If they go below freezing, the rhizomes may also be killed. Moist soil on very cold nights helps prevent rhizome death.

Wind causes abrasions and tearing of the leaves, similar to its effects on bananas and strelitzia. Windbreaks are useful for growing heliconias outdoors. Lack of boron can cause a curious scratching symptom in the foliage.

Numerous small, scattered scratches will develop on the leaves, even in the absence of any wind. Foliar sprays with Solubor (sodium borate), borax, or a boron chelate should correct the problem in the subsequent foliage.

Root blackening can be caused by root diseases, nematodes, or poor aeration and drainage. When the roots are weak in *Heliconia*, a number of the older leaves generally turn medium brown. The root problem needs to be corrected to restore plant vigor.

TRICKS

The growth regulator Bonzi (paclobutrazol) is very useful on certain *Heliconia* varieties. A spray of one-half to 1 ounce of Bonzi per gallon (15 to 30 ml per 3.8 l) will do a great deal to help keep plants compact. The technique is especially helpful when growing under shade or with varieties that tend to stretch. Remember, Bonzi is absorbed through stem tissue, so you need adequate stem exposure and plant development for Bonzi to work effectively. If you elect to apply Bonzi, do so before the plants are stretched. The growth regulator may also cause the leaves to be somewhat more round than usual.

Because of leaching from irrigation and high requirements of potassium and magnesium, topdress applications with Sul-po-mag (O-O-22-10) are very helpful in maintaining good color and bloom quality in heliconias. This works in both the landscape and container nursery production. Typical rates are about 1 teaspoon per 10-inch (25-cm) container. The good *Heliconia* grower learns to manipulate sun and shade for best leaf color and flower quality. Plants grown in high light until blooms initiate will continue to flower after shading.

INTERIOR CARE

Newly installed heliconias can bloom for three to four months at 350 f.c. (3.8 klux). Blooming may last longer under higher light. A flower can last as long as five weeks. Repeat blooming generally does not occur unless interior light levels are adequate—the more light the better.

Interior heliconias require fairly frequent water, or browning of older foliage will result. Fertilize gently but don't allow soluble salts to become high under low light conditions. When you do irrigate, water the plant thoroughly. Humidity of 40% or better is helpful, or leaf burn may increase. Groom the plants periodically, removing spent flowers and leaves. Keep an eye out for insect pests, as well, especially on newly installed plants.

REFERENCES

Ball, D. 1986. Hues of Heliconia. *Interior Landscape Industry* (August): 25-29.

Ball, D. 1987. Heliconia Update. *Nursery Digest* (August): 37-41.

Broschat, T.K., and H.M. Donselman. 1984. Growing *Heliconia psittacorum* for Cut Flowers. *Nurserymen's Digest* (June): 42-43.

Broschat, T.K., H.M. Donselman, and A.A. Will. 1984. *Andromeda, a Red and Orange Heliconia for Cut-Flower Use.* University of Florida IFAS Circular S-309, 1-5. Gainesville: University of Florida Institute of Food and Agricultural Sciences.

Donselman, H.M., and T.K. Broschat. 1987. Commercial Heliconia Production in South Florida. *Nurserymen's Digest* (January): 49-53.

Griffith, L. 1984. Heliconia and Ginger: Cut Flowers With a Future. *Florist's Review* (August 16): 36-37.

Watkins, J.V., and T.J. Sheehan. 1975. *Florida Landscape Plants,* 111. Gainesville: The University Presses of Florida.

HIBISCUS

HABITAT

Hibiscus plants have been in cultivation as ornamentals for centuries, first in the Orient, later in Europe, and in Florida for over 100 years. They are grown for both indoor and outdoor use in virtually all of the tropical and subtropical parts of the world. *Hibiscus* is an ancient Greek and Latin name. The name of the most commonly grown species, *H. rosa-sinensis*, means "Chinese rose."

Many hibiscuses, including *Rosa-sinensis*, are native to tropical Asia, though one source suggests they may actually be native to a sunken continent near Madagascar! Specifically, hibicuses tend to grow naturally in swampy areas. Because of this they generally require a steady moisture supply and are not tolerant of drying out. Their swampy habitat also explains the brittle nature of hibiscus wood and the fact that the plants are not very tolerant of

wind. They will per-
form well, however, in
poorly ventilated situ-
ations, including stag-
nant greenhouses. The
plants are not sensi-
tive to photoperiod.

USES

I used to grow 6- and
10-inch (15- and 25-
cm) hibiscuses by the
thousands for the
landscape trade when
I first started in the

Fig. 29 A well-grown 10-inch hibiscus, treated with Cycocel.
(Courtesy of Ed Clay)

business. I remember huge tractor-trailers from Canada coming into the nursery during March and April, and they would buy virtually every hibiscus I
had. For a while I wondered why Canadians would be coming in to buy tropical hibiscuses in large quantities. The drivers told me that the hibiscus flowers were popular with Canadian homeowners as signs of spring coming early
after the long Canadian winter. Once it warmed up enough, hibiscuses were
popular as patio plants during the summer, and their colorful flowers were
cherished.

Today hibiscuses are still big spring and summer sellers, with March
through May being the peak sales season. The plants are commonly grown as
flowering bushes in 5- to 10-inch (12.5- to 25-cm) containers. Standard tree
forms are also grown in larger containers, and I have seen some growers produce attractive braided standard trees in 10-inch pots. Small hedges and trees
are ubiquitous in the tropics, and hibiscuses treated with growth regulator are
occasionally seen as bedding plants in Hawaii and elsewhere.

VARIETIES

Hibiscuses come in numerous colors, including red, white, pink, orange, yellow, and multicolored flowers. Both single- and double-flowered varieties are
common, as are several odd flower forms. The nomenclature of hibiscus cul-

tivars is rather inconsistent, and the common names of these varieties may be different in Europe and elsewhere.

Good red-flowered varieties for foliage production include Brilliant, which is a classic single red variety. Moissiana is smaller, with leaves more like a grape ivy and smaller single red blooms. Single Classic Pink and Seminole Pink are good, large-flowered, single pink varieties. Seminole Pink feels somewhat like sandpaper on the leaf underside and is rather sensitive to Cycocel. Salmon, an attractive, spreading, single pinkish variety, is also very Cycocel sensitive.

Double-flowered hybrids exist in most of the colors, a popular one being Double Classic Pink. Among the single-flowered varieties, Casablanca has an attractive white flower with a red throat. Snow Queen has variegated foliage with single red flowers. The species *H. schizopetalus* has unusual fringed flowers, while *H. syriacus*, the rose of Sharon, is popular in temperate landscapes. The last three varieties mentioned are generally not grown as foliage plants.

Paint exhaust fans yellow to attract insects and blow them out of the greenhouse.

PROPAGATION

Most all hibiscus propagation is from tip or hardwood cuttings. Cuttings of crisp to hard wood may be taken anytime, rooting in four to six weeks under warm conditions, eight to 10 weeks under winter conditions. Leaf-and-eye cuttings are sometimes used, but rooting tends to be uneven. Hisbiscuses may also be propagated by grafting, budding, air-layering, and seed. Most of these measures, however, are not utilized for commercial foliage production. Even with hardwood cuttings, some cultivars are easier to root than others.

Some nurseries maintain stock plants, but many nurseries get their cuttings by trimming existing production plants. Remove any flowers or buds from the cuttings during propagation. Cuttings are typically about 4 inches (10 cm) long. Dipping the cuttings in 0.5% IBA (indolebutyric acid and talc), perhaps with Captan added as a fungicide, is beneficial. Cuttings are stuck in a fifty-fifty blend of peat and perlite or a similar open mix and usually misted for five seconds every 10 minutes. Some shade is usually placed over the misting area to reduce moisture stress in the cuttings. Also, keeping the rooting medium between 79 and 86F (26 and 30C) with bottom heat helps in winter.

CULTURE

For best flowering, hibiscuses should be grown in full sun. A few varieties can tolerate a minimum of 1,000 f.c. (10.8 klux) in northern greenhouses in winter, though a minimum of 3,500 f.c. (37.7 klux) is better for most situations. With light shading, leaf color may be improved, but you can still maintain decent bloom count. The potting mix should have good moisture-holding capacity to avoid serious drying and a pH of between 5 and 6.5.

The plant generally grows better at warm temperatures, typically 75 to 90F (23.9 to 32.2C). Growth begins to slow as temperatures drop below 70F (21.1C). Plants may be damaged or killed at 28 to 30F (minus 2.2 to minus 1.1C). Fertility status has little effect on cold tolerance.

Grower preference varies on how many cuttings to make per plant and how many times to prune plants. Many growers like to use a three-cutting liner to start a 6-inch (15-cm) pot and trim it twice. Some place as many as five cuttings per pot, though that is not common. The number of plants per pot also varies somewhat by container size and variety. Typical spacing is 12-inch centers (30-cm) for 6-inch pots. The containers can be placed pot to pot early, then spaced about the time of the first pinch. Centers of 24 inches (60 cm) are generally used for 10-inch (25-cm) pots. Plants will break better if they are cut back to at least semihard wood. The best cutback times are at approximately six and 11 weeks.

NUTRITION

Hibiscuses may be fertilized rather aggressively, as they are fairly tolerant of soluble salts, though less tolerant of ammonia toxicity. Ratios of 1-1-1 or 2-1-1 are popular. At least 7.5 to 10 grams of 14-14-14 or 13-13-13 slow-release fertilizer every three months is common for a 5-inch (12.5-cm) container. Many growers, especially those in northern climates and those growing smaller pots, prefer constant liquid feed of 20-20-20 at about 250 to 300 ppm nitrogen. An occasional leach is desirable if growing in a covered greenhouse. Better production frequently occurs if granular or slow-release fertilizer is supplemented with liquid feed. Increasing potassium in liquid feed in the latter stages of production may improve bloom count somewhat.

Hibiscuses are not susceptible to many nutritional deficiency symptoms. Lack of nitrogen will result in pale foliage and slow growth. Lack of iron is fairly common, especially when pH is high, root problems are evident, or if plants are potted too deeply. Foliar sprays of iron don't work very well. Iron is best applied on this plant as either a granular topdress or a soil drench.

TABLE 18 Leaf analysis rating standards for *Hibiscus rosa-sinensis*

Nutrient (%)	Very low	Low	Medium	High	Very high
Nitrogen	<2.0	2.0–2.4	2.5–3.0	3.1–4.5	>4.5
Sulfur	<0.13	0.13–0.19	0.20–0.50	0.51–1.00	>1.00
Phosphorus	<0.20	0.20–0.24	0.25–1.00	1.01–1.25	>1.25
Potassium	<1.25	1.25–1.49	1.50–3.00	3.01–4.50	>4.50
Magnesium	<0.21	0.21–0.24	0.25–1.00	1. 01–1.25	>1.25
Calcium	<0.85	0.85–0.99	1.00–2.00	2.01–3.00	>3.00
Sodium			<0.21	0.21–0.50	>0.50
(ppm)					
Iron	<40	40–49	50–200	201–1000	>1000
Aluminum			<250	251–2000	>2000
Manganese	<25	25–39	40–200	201–1000	>1000
Boron	<20	20–24	25–100	101–150	>150
Copper	<4	4–5	6–200	201–500	>500
Zinc	<16	16–19	20–200	201–1000	>1000

Sources: Institute of Food and Agricultural Sciences, Apopka, Florida; Dr. Benjamin Wolf, Fort Lauderdale, Florida.
Notes: Common name: hibiscus. Sample of most recent fully mature leaves, no petioles.

Molybdenum deficiency is occasionally observed, with foliage becoming blasted or strappy in appearance. As long as you have some molybdenum in the feed program, a deficiency should not occur.

DISEASES

Four bacterial diseases attack hibiscuses with some regularity. *Xanthomonas campestris* pv. *malvacearum* is common in the tropics. During warm weather older and middle foliage show large, brown to black, angular leaf spots. Affected foliage generally abscises. *Pseudomonas cichorii* is also prevalent in warm, wet weather. Older and middle leaves are affected, and the irregularly shaped lesions have black-and-purple borders. *Pseudomonas syringae* is more active in cooler temperatures. Its pinpoint leaf spots are brown to black, primarily on the new leaves, which tend to pucker. *Erwinia* is occasionally found as a cutting rot in propagation. Keep foliage dry to help combat bacterial diseases. Many growers spray with Dithane (mancozeb) plus a copper fungicide in a tank mix to control bacterial disease, though it is only somewhat effective.

Several fungal diseases are also commonly encountered. *Pythium* causes root rot when soil stays excessively wet or is highly compacted. *Phytophthora* causes similar symptoms, though roots tend to turn black, and the entire plant will die back. Drying plants out in addition to drenching with Subdue (metalaxyl) or Aliette (fosetyl-aluminum) should aid in control. *Phomopsis* may cause a fungal dieback, especially when plants have root problems or are subjected to moisture stress. Prune out dead twigs with sterilized clippers, then spray with a thiophanate methyl fungicide, such as Cleary's 3336 or Domain.

A few viruses are known to attack hibiscuses, though they are relatively rare and not particularly serious. Root knot nematodes can cause problems in landscape situations, though they are rarely seen in foliage production.

INSECT AND MITE PESTS

Whiteflies can be a particularly severe problem, and controlling them can be a significant production cost. Growers in the tropics frequently tank-mix a whitefly insecticide with their Dithane (mancozeb) and copper spray. Some common chemicals for whitefly include Thiodan (endosulfan), Talstar (bifenthrin), Azatin (azadirachtin), insecticidal soaps, and horticultural oils. Precision (fenoxycarb) is a useful insect growth regulator against whiteflies, and yellow sticky traps are also beneficial. The neem extract product Naturalis-O is another useful product.

Scale insects are also troubling at times on hibiscuses. One curious scale insect rarely seen on other plants is the hand grenade scale. This tiny, light brown scale insect, which tends to resemble a hand grenade, is frequently found on stems near where the leaves emerge. Its sucking activity causes reduced vigor, defoliation, and ultimately plant death. This scale is very difficult to spot unless you know what you are looking for. Snow scale is frequently found on bark, especially in the lower part of the plant. This scale is about a millimeter long, rectangular, and bright white. Applications of granular Di-syston are effective on scale pests, as are sprays of Cygon (dimethoate) and horticultural oils.

Aphids occasionally attack new growth, especially in spring. Numerous chemicals are registered for aphids, including Orthene (acephate), Knoxout (diazinon), Tempo (baythroid), and insecticidal soaps. Two-spotted mite infestations are common, and occasionally you will see European red mites in addition to predator mites *(Phytoseiulus)*. Options include insecticidal soaps, horticultural oil, Avid (abamectin), and Pentac (dienoclor).

DISORDERS

Iron chlorosis occurs as distinct yellowing in much of the plant, usually with green veins visible. Factors contributing to iron chlorosis include excessively wet or compact soil, root rot disease, nematodes, or elevated pH. Try to correct the fundamental problem, and then drench or top-dress with chelated iron or iron sulfate. If plants run too dry for even a day, older leaves will suddenly turn canary yellow a couple of days later, then drop. With good moisture, regrowth of foliage will occur at these sites in about six weeks.

Bud drop in hibiscuses has many potential causes, including hot, dry, or windy weather; thrips infestation; nematodes; or dark storage. Bud drop is ultimately caused by ethylene generation within the plant. *Botrytis* attacks flowers, especially spent blooms in cool, wet weather. Nematode infestation causes hibiscuses to be twiggy, thin, lacking in vigor, and with few flowers. Knots may or may not be visible on the roots.

If hibiscuses are grown too close together, leaf drop may occur from ethylene generation in wet weather. Poor flowering on the sides and lower portions of the plant will also be observed.

Most pesticides burn hibiscus flowers. The blooms only last one day whether they are on the plant or off, so they are generally quickly replaced. Malathion is usually phytotoxic to hibiscuses. Cycocel yellows can occur at times, especially when spraying in hot weather. Portions of the leaf turn yellow or golden in color after spraying with Cycocel (chlormequat). The symptoms usually disappear completely in about two weeks. The algaecide Agribrom can be injurious at 55 ppm, though 25 ppm is usually okay.

TRICKS

Hard-to-root cultivars may propagate more easily if the stock plants are grown under 47% shade. High intensity lighting is useful for producing quality stock and production hibiscuses during northern winters. Sprays of potassium nitrate at $1\frac{1}{2}$ pounds per 100 gallons (0.7 kg per 400 l) help improve bloom count slightly in most situations. In propagation, trimming back part of the large basal leaves prior to sticking cuttings will help reduce disease incidence under mist.

Don't fertilize the variegated variety Snow Queen aggressively, or the plants will tend to revert to the nonvariegated form. In order to reduce hibiscus bud drop during shipping, a quick spritz on the inside of the sleeves with 2.0 mM silver thiosulfate will help reduce ethylene generation and bud drop.

Sprays of Cycocel (chlormequat) are very effective in improving this plant's quality. One or two Cycocel sprays generally result in greater leaf color and shine, a more compact growth habit, more blooms, and more even flowering. The technique has been known since the 1960s, and although it is not essential, Cycocel helps a great deal in creating a florist's-quality hibiscus plant. Rates and timing depend on plant size and cultivar. Sprays, typically of 0.5 to 1 ounce per gallon (15 to 30 ml per 3.8 l), are usually applied two to four weeks after plants are cut back, when new growth is 1 to 1.5 inches (2.5 to 3.75 cm) long. The spray is frequently applied at production week 12 to 13 for 6-inch (15-cm) pots and on some varieties is repeated two weeks later. More intense flowering will be observed six to 11 weeks later, or about eight to 10 weeks in Florida. The Cycocel effect carries over in the plant even after pruning. An unexpected benefit from the growth regulator spray is a greatly reduced incidence of bacterial leaf spot.

INTERIOR CARE

Hibiscus plants are frequently kept outdoors on patios during summer months in the north, then brought in for protection during the winter. If you hope to get any flowers at all in the interior, give the plants as much light as you can. Because interior blooming is fairly sparse unless light is very high, conservatively fertilize to avoid weak, stretched growth.

Hibiscus plants will require regular watering indoors; avoid letting them dry excessively. Watch for spider mite infestations. The minimum temperature should be no lower than 68F (20C) for best plant quality. Plants previously treated with Cycocel tend to remain more compact in the interior environment.

 The primary function of molybdenum is to help convert ammonia to nitrate.

REFERENCES

Aimone, T. 1985. Culture notes. Hibiscus (Rose of China, China Rose), Genus, species: Hibiscus rosa-sinensis, Family: Malvaceae. *GrowerTalks* (September): 26-29.

Carpenter, W.J. 1992. Hibiscus Cuttings Need Warm Medium. *Florida Nurseryman* (November): 38-40.

Chase, A.R. 1986. Three Bacterial Leaf Spots of *Hibiscus rosa-sinensis*. *Nurserymen's Digest* (July): 70-71.

Chase, A.R. 1988. Cycocel Reduces Severity of Bacterial Diseases on Hibiscus. *Foliage Digest* 11 (5): 1-3.

Chase, A.R., and R.T. Poole. 1988. *Effects of Fertilizer Rates on Growth of Hibiscus rosa-sinensis 'Brilliant Red'.* CFREC-Apopka Research Report RH-88-2: 1-3. Apopka: Central Florida Research and Education Center.

Criley, R.A. 1980. Potted Flowering Hibiscus. *Florist's Review* (February 21): 48-49, 64.

Force, A.R., K.A. Lawton, and W.R. Woodson. 1988. Dark-Induced Abscission of Hibiscus Flower Buds. *Hortscience* 23 (3): 592-593.

Golby, E.V., ed. 1978. *What Every Hibiscus Grower Should Know,* 76-77. Pompano Beach, Fla.: The American Hibiscus Society.

Goode, T.L. Summer 1996. Telephone conversation with the author.

Griffith, L.P. 1983. Technology & Trends. *Florist's Review* (June 9): 29-30.

Miller, R.O. 1987. Take a Look at Hibiscus. *GrowerTalks* 51 (4): 136-140.

Teets, T.M., and R.L. Hummel. 1985. Nutritional Effects on Cold Acclimation of *Hibiscus syriacus. Proc. Fla. State Hort. Soc.* 98: 90-92.

HOWEA

Kentia Palm

HABITAT

The kentia palm, *Howea forsterana*, is one of the most elegant interior palms. The genus is named after its native habitat, Lord Howe Island, off the eastern coast of Australia. The palm was originally placed in the genus *Kentia*, and like with areca palms, the old genus name continues as the common name. Other common names include "thatch palm," "paradise palm," and "sentry palm." *Howea belmoreana* is a similar palm, though the leaflets are more erect and upright in mature specimens, and it generally grows at higher elevations in its native habitat. *H. forsterana* is somewhat faster growing than *H. belmoreana* and is therefore preferred in the foliage trade.

196

Kentias are native to lowland forests on Lord Howe Island, usually no higher than 500 feet (152 m) above sea level. They are usually found in sandy soils. Climate on Lord Howe Island is warm to cool temperate. Kentias are therefore tolerant of wind, drafts, temperature changes, and rough handling. They take cool weather well but cannot tolerate frost. Kentias frequently grow on slopes in their habitat, so standing water is undesirable elsewhere.

USES

Kentias have been popular indoor palms since the Victorian era. The 1887 book *Practical Floriculture* lists the

Fig. 30 *Howea forsterana*, the kentia palm, in Homestead, Florida. (Courtesy of Ed Clay)

kentia as one of the best indoor palms. It is tall and slender, with arching petioles and gracefully trailing leaflets.

The most common type of production is three plants per pot in larger containers, ranging from 10 to 17 inches (25 to 42.5 cm). Grown this way the palms make an attractive, tall, full specimen. Kentias are also attractive grown as single palms, especially in very large tubs. Because of their height kentias are frequently installed as focal points in interior landscapes.

VARIETIES

H. forsterana is the preferred kentia in the trade today. It is not particularly fast growing, but it is more rapid than its counterpart, *H. belmoreana*. In commercial plantings of *H. forsterana*, it is common to see some of the young palms with reddish petioles. These are typically referred to as red kentias, which growers usually pot together. The red petiole effect tends to become less noticeable as the plants mature.

PROPAGATION

Like virtually all palms, kentias are propagated from seed. Most of the seed comes from Lord Howe Island, though some is also harvested from mature specimens in Australia and California. Demand for kentia seed generally exceeds supply. The seeds range from orange-brown to red when mature. It takes between two and four years from the time of flower opening for a kentia seed to be ready to be harvested. This, combined with the fact that it generally takes 15 years for a young kentia to be able to flower, should ensure a limited availability of these palms in the future—and therefore a good price.

Seeds are generally planted shallowly, just barely covered or partly visible. A fifty-fifty blend of peat and sand makes a desirable propagation medium, though straight, reasonably coarse sand is also good. Germination is irregular, ranging from two months to one year. Bottom heat is helpful in speeding germination, as with *Chamaedorea*. If you are using bottom heat, make sure you monitor the moisture levels in the seed bed frequently, or excessive drying may kill germinating seeds.

CULTURE

Kentia palms are quite slow growing. It may take as long as eight years to get a 3-foot (0.9-m) plant, though most growers are producing kentias much faster today. The best light levels range from 2,500 to 6,000 f.c. (27 to 64.8 klux). A fifty-fifty mixture of peat and sand or peat and perlite is preferred because you don't want too high a moisture-holding capacity for the mix. Try to have at least 40% of the medium comprised of components that won't physically break down over time. Too many growers in Florida try to produce kentias in heavy peat-sawdust blends, which break down and contribute to plant loss. California growers tend to use much sandier mixes. The desired pH range is 6 to 6.5.

Most growers transplant seedlings into 6-inch (15-cm) individual pots. Transplant the seedlings from the seed bed before the roots become too extensive, and don't pot them deep in the 6-inch pot, or losses will occur. When the young plants are about 10 to 12 inches (25 to 30 cm) high, they can be transferred to larger containers. Seedling vigor may be variable. Some growers like to put the aggressive growers together in one larger pot, while others like to mix and match to end up with plants of different heights in the container.

The preferred kentia temperature range is 60 to 85F (16 to 29C). Growth begins to slow below 60F (16C), though the plants easily tolerate 50F (10C). Temperatures usually need to approach freezing before significant cold damage will occur, though when soils are cold, kentias virtually do not grow.

NUTRITION

Because of the slow rate of growth in this palm, gentle, long-term, slow-release, coated fertilizers are popular. Ratios of 3-1-2 of N-P$_2$O$_5$-K$_2$O are usually selected, preferably with adequate magnesium. The slow-release fertilizers can be top-dressed or incorporated into the medium. Some growers prefer to gently feed the seed beds and the 6-inch (15-cm) liners with liquid feed, usually 150 ppm nitrogen from 20-20-20, once a week.

Magnesium deficiency is common in older kentias. The older leaves gradually become a golden yellow color, progressing from the leaflet tips inward. Adequate soil magnesium levels and a calcium-magnesium ratio in the soil of less than eight to one usually prevents this disorder. Potassium deficiency appears as small golden to tan spots in the older foliage. Once kentia leaves become deficient in potassium or magnesium, an affected leaf will generally not recover and has to be removed. It is therefore best to prevent such disorders. Zinc deficiency results in smaller than normal leaves, which have a somewhat stubby appearance. Lack of manganese can cause a kind of leaf distortion known as frizzle-top, where foliage is severely withered and distorted.

DISEASES

Probably the most serious kentia disease is *Phytophthora*, which means "plant destroyer" in Latin. If moisture-holding capacity in the medium is too high or if plants are overwatered, kentias frequently die from *Phytophthora*. Occasional applications of Subdue (metalaxyl) or Aliette (fosetyl-aluminum) help prevent *Phytophthora*.

The fungus *Colletotrichum* causes fairly large, roundish, tan leaf spots. As with all leaf spots, keeping foliage dry is helpful. Sprays of Dithane (mancozeb) or a thiophanate methyl fungicide, especially when applied to the underside of the foliage, also helps. *Cylindrocladium*, another common kentia leaf spot, is somewhat darker in appearance than *Colletotrichum*. Sprays of

Chipco 26019 (iprodione) are beneficial, as are cultural controls. *Helminthosporium* occasionally causes a severe leaf spot problem in shade-houses. The spots are medium to dark brown and almost perfectly round. Regular sprays of Dithane or Chipco 26019 should help, but again, you need to keep the foliage dry.

 Magnesium is in the middle of the chlorophyll molecule.

Root disease can also be caused by the fungi *Pythium* and *Rhizoctonia*. It is very difficult to tell root diseases apart visually. The grower should inspect kentia roots on a regular basis and have disease diagnoses performed if significant root loss occurs. *Rhizoctonia* can be controlled with soil drenches of Chipco 26019 (iprodione) or Terraclor (PCNB). *Pythium* control is similar to that of *Phytophthora*.

INSECT AND MITE PESTS

Florida red scale is seen fairly regularly on kentia palms. These round scales are usually no more than one-eighth inch (3 mm) in diameter; younger scales are smaller. Red scales are found primarily on the undersides of older leaves, where they create yellow leaf spots from their feeding activities. Sprays of Cygon (dimethoate) or Supracide (methidathion) control them, though several sprays will be needed. Caterpillars chew holes in kentia leaves occasionally. Use sprays of Orthene (acephate) or Sevin (carbaryl) to control them.

Spider mites, especially the two-spotted mite, attack kentia palms. The infestations are more severe in hot, dry weather. Leaf speckling is visible, and the foliage turns a rather dull gray-green color. The grower should inspect for mites on a regular basis. Controls include predator mites *(Phytoseiulus)*, insecticidal soaps, and the normal range of chemical miticides. Cygon (dimethoate) controls spider mites, in addition to scale.

DISORDERS

When soluble salts are elevated in kentias, the plants appear weak, with excessively drooping foliage. Susceptibility to disease also increases. Therefore, watch the rates on slow-release or liquid fertilizers. When salts are high, kentias can be difficult to leach, as high soil moisture from leaching can also induce disease. The best preventions are a well-drained mix with relatively low moisture-holding capacity and a conservative fertilizer regime.

Older kentia leaves frequently appear to be diseased. They are discolored and have various golden to gray speckles in the older foliage. While this looks like a disease, it is usually a combination of potassium and magnesium deficiency, possibly with some old mite injury. Fungicide sprays are usually not needed in this situation. Remove the worst old leaves, then maintain adequate potassium and magnesium fertility.

Howea is sensitive to atmospheric fluoride. Symptoms include chlorosis and marginal necrosis of leaves as well as tip burn. Palms don't appear to be especially sensitive to fluorides in fertilizers or irrigation water, but leaf-tip burn from those fluoride sources is occasionally observed. In the native kentia habitat, rats frequently raid plantations and eat the developing seeds, reducing the harvest.

TRICKS

A higher-than-normal percentage of sand or perlite in the mix is helpful in producing high-quality kentias while reducing plant losses from overwatering. In older plants the pH may drop; therefore, top-dressing with dolomite can help raise the pH and supply adequate magnesium. It also helps to occasionally supplement with a topdress of Sul-po-mag (0-0-22-10) to avoid deficiency problems with potassium and magnesium. Don't use more than 1 tablespoon per 10-inch (25-cm) container. If salts are high in older plants with decomposed media, you can reduce salts by irrigating with a sugar-and-water solution, instead of leaching. The table sugar stimulates microbial activity and helps reduce the salts without risk of applying excessive moisture.

While kentias are tolerant of wind, try to avoid wind immediately after stepping up into larger containers. Winds during this stage cause foliar burn from desiccation and may result in nonvertical growth.

INTERIOR CARE

Kentias are not tolerant of standing water indoors, so either avoid using a saucer under the container or else place stones between the bottom of the pot and the saucer to keep the plants out of standing water. Maintain moderate but not wet soil moisture. Overwatering is the major cause of death for interior kentias, so be conservative. Fertilize gently every two to three months, using about a teaspoon of soluble 20-20-20 plus one-half teaspoon of Epsom salts per gallon (3.8 l). Do not fertilize frequently, or salts will accumulate.

Indoor kentias can take very low light levels, down to 50 f.c. (0.5 klux). However, 100 f.c. (1.1 klux) or more is better. If you must keep kentias under very low light conditions, occasionally returning them to the greenhouse revitalizes them, but place them in no more than 2,500 f.c. (27 klux) of light. Under the higher light gradually increase irrigation frequency and groom the plants, if necessary.

REFERENCES

Blessington, T.M., and P.C. Collins. 1993. *Foliage Plants: Prolonging Quality*, 135-137. Batavia, Ill.: Ball Publishing.

Furuta, T. 1982. Paradise Palms Make a Comeback With Plantscapers. *Florist's Review* (January 7): 18, 52.

Stewart, L. 1994. *A Guide to Palms and Cycads of the World*, 107-108. Sydney, Australia: Angus & Robertson.

JANET CRAIG

(*See* Dracaena deremensis)

KENTIA PALM

(*See* Howea)

LIPSTICK PLANT

(*See Other Foliage:* Aeschynanthus)

MAJESTY PALM

(*See* Ravenea)

MARANTA

Prayer Plant

HABITAT

Marantas are frequently called prayer plants because their leaves tend to fold up at night like a pair of hands. The genus is named after B. Maranta, an Italian botanist. Two varieties make up the majority of foliage production. Both are strains of *Maranta leuconeura,* the species name meaning "white vein." About 20 species exist, the principal ones native to Brazil. *Maranta*

Fig. 31 Four-inch red maranta, Apopka, Florida. (Courtesy of Marshall Horsman)

is a fairly close relative of *Calathea* as well as the West Indian arrowroot.

Marantas are generally found in sheltered, shaded areas in their native habitat. They seem to prefer locations free of high winds and cold temperatures. The herbaceous stems grow along the ground and sometimes up slopes and over rocks. There are no tendrils in the stems, so they have a very limited ability to climb.

USES

Most marantas are produced in small pots, with sizes between 3 and 6 inches (7.5 and 15 cm) making up the majority of commercial production. Small hanging baskets are also grown, with pot sizes typically between 5 and 8 inches (12.5 and 20 cm). Marantas do not grow tall, so larger containers tend to appear way out of proportion.

Prayer plants are ubiquitous in dish garden production and other types of combinations. They are frequently planted in interior landscapes as ground covers, and they may also be used to trail over ledges or window boxes. The plants perform well over long periods in reduced light situations.

VARIETIES

Two principal cultivars make up the vast majority of maranta production. *Maranta leuconeura* var. *erythroneura* is commonly known as red maranta. The common name "red nerve" plant is also sometimes used, as *erythroneura* means "red nerve." Red maranta has bright red leaf veins, with a greenish yellow center portion of the leaf and a green to black outer background. The foliage is reddish purple underneath.

Maranta leuconeura var. *kerchoviana* is most commonly known as green maranta. "Rabbit's foot" and "rabbit's tracks" are other common names, referring to the two rows of five black spots on the leaves, which somewhat resemble rabbit's tracks. When mature it grows in a clumping fashion, with stems also growing along the ground. The leaves are green and satiny on the upper surface, with regularly arranged, large, black spots. Like red maranta, green maranta frequently produces small, white flowers.

A relatively rare cultivar is *Maranta leuconeura* var. *leuconeura,* the silver feather maranta. The foliage is a curious mix of grayish blue-green. Silver Feather is an attractive plant that has not really caught on in the U. S. trade.

PROPAGATION

Marantas are propagated commercially via cuttings or tissue-culture liners. Tissue-culture marantas are popular and tend to be more free of disease. Many nurseries in Central Florida and elsewhere maintain stock beds of maranta for cutting. In order to save space, stock may otherwise be maintained under groups of hanging baskets or under benches containing other plants. Cutting yield, however, is greater under 60% shade.

Cuttings containing one to two nodes with three to four leaves may be taken anytime, though cutting yield is about 40% greater in summer. Two cuttings are frequently planted in a 3-inch (7.5-cm) pot, three cuttings in a 4-inch

(10-cm) pot. Under warm conditions they may be misted once or twice a day. Generally, though, mist is discouraged in order to avoid disease problems.

CULTURE

Production in covered greenhouses is preferred for marantas, as they do not appreciate wide fluctuations in temperature or moisture. A greenhouse maintained between 70 and 80F (21 and 27C) is ideal. Try to avoid temperatures above 90F (32C) or below 50F (10C). Light levels depend somewhat on temperature, but the normal range is 1,000 to 2,500 f.c. (10.8 to 26.9 klux). Shadehouse production is discouraged in most areas.

A relatively absorbent, peat-based mix with a pH of 5.5 to 6.5 is usually preferred. Stay toward the higher end of that pH range if you have irrigation water containing more than 0.2 ppm fluoride.

Since marantas are not very large plants, they can be grown rather close together on the bench. Some growers set the plants pot to pot first for rooting and early growth, then space them later. In some situations, though, additional space will not be required. A steady supply of moisture is helpful, but avoid extremes of wet or dry. In my experience it is best to irrigate marantas between 10 A.M. and noon.

NUTRITION

Being rather sensitive, slow-growing plants, marantas prefer gentle fertilization rates. Liquid feed of about 150 ppm nitrogen from a 3-1-2 ratio of fertilizer is common. This can come from soluble 24-8-16 or from bulk liquid 9-3-6. Some growers also use soluble 20-10-20. Because of the sensitivity of the foliage, liquid feed should be rinsed off at the end of the irrigation cycle, or foliar injury can result. Incorporating trace elements into the potting medium is advisable, but keep the rates conservative. Maranta stock beds require supplemental magnesium, as the regular harvesting of cuttings removes substantial amounts of magnesium reserves.

DISEASES

The most troublesome disease of maranta is *Helminthosporium* leaf spot, which is especially prevalent in stock beds. Spots begin as small, water-soaked lesions, then become yellow and finally necrotic. The spots are typically one-sixteenth inch (1.5 mm) wide, though under severe pressure they

TABLE 19 Leaf analysis rating standards for *Maranta kerchoviana*

Nutrient (%)	Very low	Low	Medium	High	Very high
Nitrogen	<1.50	1.50–1.99	2.00–3.00	3.01–4.50	>4.50
Sulfur	<0.15	0.15–0.19	0.20–0.50	0.51–1.00	>1.00
Phosphorus	<0.15	0.15–0.19	0.20–0.50	0.51–1.00	>1.00
Potassium	<2.25	2.25–2.99	3.00–5.50	5.51–6.50	>6.50
Magnesium	<0.17	0.17–0.19	0.20–1.00	1.01–1.25	>1.25
Calcium	<0.40	0.40–0.59	0.60–1.50	1.51–2.50	>2.50
Sodium			<0.20	0.20–0.50	>0.50
(ppm)					
Iron	<40	40–49	50–300	301–1000	>1000
Aluminum			<250	251–2000	>2000
Manganese	<40	40–49	50–200	201–1000	>1000
Boron	<20	20–24	25–50	51–75	>75
Copper	<4	4–7	8–50	51–200	>200
Zinc	<15	15–19	200–200	201–1000	>1000

Sources: Institute of Food and Agricultural Sciences, Apopka, Florida; Dr. Benjamin Wolf. Fort Lauderdale, Florida.
Notes: Common names include maranta, red-nerve plant. Sample of most recent fully mature leaves, no petioles.

may coalesce to form large, tan blotches with yellow halos. Leaves need to be wet for about six hours for the *Helminthosporium* fungus to sporulate, hence the recommendation for watering toward the middle of the day. Air movement also helps a great deal in getting the foliage dry after irrigation. Sprays of Daconil (chlorothalonil) also help control the fungus.

Cucumber mosaic virus is occasionally a problem. A mosaic pattern of yellow blotches appears on the leaves of both red and green marantas. Virus-infected leaves may be slightly distorted and smaller than normal. There are no controls other than to remove and discard infected plants and not to propagate from virus-infected stock.

Both root knot and burrowing nematodes can be severe, occasionally even in greenhouse situations. The plants are severely stunted and weak, and cutting yield will be very poor. Root growth is poor, and roots may be blackened or have abnormal swelling. No chemical controls are currently available. Remove and discard affected plant material and the soil.

INSECT AND MITE PESTS

Caterpillar injury is fairly common on marantas. The worms chew holes in the leaves, and if they do so as the leaves are unfurling, a regularly spaced series of holes in the leaves will result. Large droppings may be observed near the injured tissue, and old caterpillar damage has a somewhat more callused appearance. Excluding moths and butterflies from the greenhouse is helpful, as are sprays of Dipel *(Bacillus thuringiensis)*, Orthene (acephate), or Dursban (chlorpyrifos).

Mealybugs occasionally appear as cottony masses on the leaf axils and the lower leaf surfaces. Control can usually be achieved with repeated sprays of either Dycarb (bendiocarb), Dursban (chlorpyrifos), or Enstar (S. kinoprene).

Thrips are attracted to light blue and pink.

Spider mites, especially the microscopic mite *Steneotarsonemus furcatus*, can be very troublesome. The microscopic mite injury creates symptoms that look very much like a disease. Water-soaked lesions form on the emerging, unfurled leaves. The stems stop growing, and the plants can ultimately die. Sprays of Thiodan (endosulfan) should help control mites, especially when combined with a surfactant. Two-spotted mites occasionally attack, causing the normal leaf-speckling symptoms. Sprays of Kelthane (dicofol) or Pentac (dienoclor) are helpful.

Scale insects are not very common on maranta, but treat as for mealybugs if they occur. Snails and slugs frequently feed on maranta stock beds at night. Their chewing injury is more irregular in size and shape than that of caterpillars. The droppings are also smaller and somewhat squiggly in appearance, rather than round. Metaldehyde baits may help some, as may sprays of Grand Slam (mesurol).

DISORDERS

Scorch of leaf tips is probably the most common problem in maranta production. The distal edge of the leaf becomes medium tan in color, frequently with a yellow halo. Tip scorch is normally caused either by elevated soluble salts in the medium or by accumulation of fluorides. Monitor salts carefully, keep calcium levels up, and do your best to use low-fluoride fertilizers and irrigation water. Atmospheric fluoride injury is fairly rare in most areas. The red maranta, especially, shows irregular necrotic splotching, and faded colors. Iron chlorosis, which results in a yellowing of the newer foliage, is also fairly common. Gentle sprays with chelated iron or a soil drench with an iron product usually corrects the problem.

Growth will be greatly diminished if daytime greenhouse temperatures exceed 90F (32C). Cold injury occurs at about 45F (7C). The foliage then becomes more distinctly red than normal, with significant amounts of leaf necrosis.

TRICKS

It helps to have a little ammonical nitrogen in the liquid feed for maranta plants, primarily for acidity. Plants grown with all-nitrate nitrogen are somewhat more likely to develop iron chlorosis. Keep ammonium nitrogen at 25% or better for most situations. Incorporating gypsum into the potting soil helps maintain a steady pH, and the additional calcium helps tie up fluoride. A typical rate would be about 3 pounds per cubic yard (1.8 kg per m³). Spray stock plants regularly, at least once a month, with magnesium sulfate at 2 pounds per 100 gallons (0.9 kg per 400 l).

Prayer plants respond well to humidity and gentle air movement in the greenhouse. The humidity helps reduce tip-burn problems, and air movement helps reduce disease incidence.

INTERIOR CARE

Prayer plants do well in a consistent interior environment. Indirect or curtain-filtered light is preferable, but marantas tolerate low light better than many other foliage plants. Interior light should be at least 75 to 150 f.c. (0.8 to 1.6 klux). Bright light tends to burn the plants. Keep temperatures mild, generally between 65 and 85F (18 and 29C).

I prefer to grow marantas on the slightly dry side in the interior, especially if humidity is high. Under low-humidity situations, leaf burn increases. Keep a pan of moistened gravel near the plants, if possible, when maintaining marantas in low-humidity situations. Don't fertilize unless you have reasonably good light levels.

REFERENCES

Bailey, L.H., and E.Z. Bailey. 1976. *Hortus Third*, 712-713. New York: Macmillan.

Blessington, T.M., and P.C. Collins. 1993. *Foliage Plants: Prolonging Quality*, 141-143. Batavia, Ill.: Ball Publishing.

Chase, A.R., R.W. Henley, and L.S. Osborne. 1985. *Marantas*. AREC-A Foliage Plant Research Note RH 85-8: 1-3. Apopka: Agricultural Research and Education Center.

Conover, C.A., R.T. Poole, and C.A. Robinson. 1981. Influence of Dolomite, Fertilizer Level and Season on Stock Production of Prayer Plant, *Maranta leuconeura* var. *kerchoviana*. *Foliage Digest* (May): 15-16.

NEPHTHYTIS

(See Syngonium)

NERVE PLANT

(See Other Foliage: Fittonia*)*

NORFOLK ISLAND PINE

(See Araucaria)

PARLOR PALM

(See Chamaedorea)

PEACE LILY

(See Spathiphyllum)

PEPEROMIA
Baby Rubber Plant

HABITAT

Peperomias have been popular as small houseplants for many years. Of about 1,000 species that exist, approximately 100 varieties are cultivated, with about 10 varieties making up the majority of commercial peperomia production. Many cultivated peperomias are varieties of the species *Peperomia obtusifolia*. The name *Peperomia* refers to the pepperlike foliage, while *obtusifolia* refers to the oval leaf shape. The different varieties have different common names, though they are sometimes referred to as a group as baby rubber plants.

Many peperomias grow as epiphytes in their native habitats, which are usually in the tropics of the Western Hemisphere. *P. obtusifolia,* for example, is native to tropical America and South Florida. Because of their habitat and epiphytic tendencies, peperomias like humidity, low soil moisture, and low light levels. Like many epiphytes, peperomias are rather easy to root.

USES

Almost all peperomia production is in small containers, typically 2.5 to 6 inches (6 to 15 cm). The most popular container size for peperomias is 3 inches (7.5 cm), this is partly because the relatively small peperomia plant will be in proportion to its container. The

Fig. 32 Variegated and green peperomia production in Apopka, Florida. (Courtesy of Marshall Horsman)

DISEASE, PEST, AND DISORDER GUIDE

AGLAONEMA

Xanthomonas blight of *Aglaonema* Maria shows the characteristic bronze scorch of older leaves.

(COURTESY OF JOHN GATTI)

AGLAONEMA

Colletotrichum leaf spot of *Aglaonema* Silver Queen: The large blotches frequently appear shortly after cuttings are stuck.

(COURTESY OF OLAF RIBIERO)

AGLAONEMA

Fusarium stem rot of *Aglaonema* Maria: This fungus often causes basal rot that prevents cuttings from rooting.

(COURTESY OF JOHN GATTI)

ANTHURIUM

Pythium root rot can cause serious losses in anthuriums, despite growing on inverted pots. Root rot is a frequent occurrence when growing anthuriums in shade houses.

(COURTESY OF ED CLAY)

ANTHURIUM

Anthuriums with severe foliar marginal blight caused by *Xanthomonas*. This difficult to control bacterial disease can cause huge losses.

(COURTESY OF MARSHALL HORSMAN)

ANTHURIUM

This cold-damaged *Anthurium* Lady Jane was exposed to temperatures close to freezing. The leaves turn golden yellow and droop.

(COURTESY OF JOHN GATTI)

ARAUCARIA

Colletotrichum needle blight of Norfolk Island pine causes numerous leaf tips to turn brown, especially in dense plantings and windy conditions.

(COURTESY OF MARSHALL HORSMAN)

CHAMAEDOREA

Gliocladium stem rot of chamaedorea palm: This pathogen prefers cooler temperatures, and can be primary or secondary.

(COURTESY OF OLAF RIBIERO)

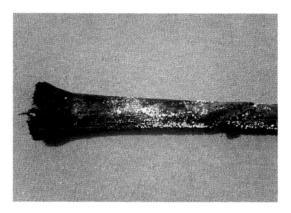

CODIAEUM

Gleosporium leaf spot on Croton Gold Dust is a common problem during rainy tropical summers.

(COURTESY OF OLAF RIBIERO)

CORDYLINE

Fluoride toxicity can cause tip burn in *Cordyline* Red Sister. Cordylines are among the most fluoride sensitive of foliage plants.

(COURTESY OF OLAF RIBIERO)

DIEFFENBACHIA

Colletotrichum blight of *Dieffenbachia:* The large brown blotches often have black and white fruiting bodies of the fungus visible on the undersides of the leaves.

(COURTESY OF OLAF RIBIERO)

DIEFFENBACHIA

Erwinia petiole rot of *Dieffenbachia:* This foul-smelling rot is often spread by cutting tools.

(COURTESY OF JOHN GATTI)

DRACAENA

Notching of *Dracaena deremensis* cv. Warneckii is due to boron deficiency. The slits in the leaves are more common under high light and high temperatures.

(PHOTO BY THE AUTHOR)

DRACAENA

Fluoride injury to taller *D. fragrans* cv. Massangeana cane: The heads on taller canes show tip burn, while the more shaded lower heads do not.

(PHOTO BY THE AUTHOR)

DRACAENA

Fusarium leaf spot of *Dracaena reflexa* Song of Jamaica: The orange border is characteristic of this fungus.

(COURTESY OF JOHN GATTI)

EPIPREMNUM

Southern blight of pothos, caused by *Sclerotium rolfsii:* The disease attacks a number of foliage plant varieties, usually in hot weather.

(COURTESY OF JOHN GATTI)

EPIPREMNUM

Pythium root rot of pothos: Roots turn black when plants are overwatered.

(COURTESY OF JOHN GATTI)

FERNS

Leaf distortion from excess nitrogen in bird's-nest fern: This variety must be fertilized gently to avoid distortion.

(COURTESY OF JOHN GATTI)

FERNS

Rhizoctonia blight of Boston fern: Fronds turn black and drop, often very quickly in a warm greenhouse.

(COURTESY OF MARSHALL HORSMAN)

FICUS

Cuban laurel thrips damage of *Ficus retusa* Nitida: The insect attacks this species of ficus, but few others.

(COURTESY OF DOUG PALMER)

FICUS

Physiological disorder of *Ficus benjamina,* possibly caused by sulfide toxicity, is often seen after heavy rains. The leaves drop, but the plant quickly recovers.

(COURTESY OF JOHN GATTI)

FICUS

Myrothecium leaf spot of *Ficus decora:* Notice the circular shape of the spot and how wet the foliage is.

(COURTESY OF JOHN GATTI)

FICUS

Ethylene-induced leaf drop of *Ficus decora* plants that were grown too close together: Lack of space prevents the ethylene gas from dissipating.

(PHOTO BY THE AUTHOR)

GARDENIA

Phytophthora blight of *Gardenia:* This black rot is usually encountered under wet conditions.

(COURTESY OF JOHN GATTI)

HIBISCUS

Iron chlorosis of *Hibiscus* is usually associated with high soil pH, root rot, or nematodes.

(COURTESY OF JOHN GATTI)

HIBISCUS

Bacterial leaf spot of hibiscus plant caused by *Xanthomonas:* The spots are angular or jagged, with a water-soaked appearance.

(PHOTO BY THE AUTHOR)

HIBISCUS

Edema on this hibiscus is caused by wide fluctuations in moisture levels. Raised brown blotches appear on the underside of the leaf.

(COURTESY OF JOHN GATTI)

HOWEA

Potassium deficiency of kentia palm affects older leaves and is frequently mistaken for a fungus disease. This problem is common in interior plantings.

(COURTESY OF ED CLAY)

MARANTA

Red maranta stock beds growing in Guatemala show magnesium deficiency symptoms. Constant harvesting of cuttings removes magnesium reserves in the stock.

(PHOTO BY THE AUTHOR)

PHILODENDRON

Manganese toxicity of *Philodendron* Black Cardinal, while not very common, is usually caused by low pH.

(PHOTO BY THE AUTHOR)

PHILODENDRON

Spots from spider mite injury on *Philodendron* Black Cardinal: These tan spots on older foliage are frequently mistaken for a disease.

(PHOTO BY THE AUTHOR)

PHILODENDRON

Manganese toxicity of juvenile *Philodendron selloum:* Dark purple to black specks form in abundance within the leaf.

(COURTESY OF JOHN GATTI)

PHOENIX

Oleander scale infestation of *Phoenix roebelenii:* Also called magnolia scale, this insect can infest numerous palm species, including areca and majesty palms.

(COURTESY OF ED CLAY)

RAVENEA

Classic iron deficiency of *Ravenea rivularis* is shown by mild to severe yellowing of newer leaves, especially when palms are growing rapidly.

(PHOTO BY THE AUTHOR)

SCHEFFLERA

Alternaria leaf spot and blight of *Schefflera:* This disease is most common on *S. actinophylla*, less common on the Amate variety and *S. arboricola*

(PHOTO BY THE AUTHOR)

SCHEFFLERA

Leaf drop caused by exposure of scheffleras to ethylene: Heavy rains and close plant spacing created this problem.

(PHOTO BY THE AUTHOR)

SCHEFFLERA

Edema in *Schefflera arboricola:* Excess moisture exudes through the leaves leaving these blisters behind.

(PHOTO BY THE AUTHOR)

SCHEFFLERA

Sphaerobolus popcorn fungus of *S. arboricola:* This fungus, usually mistaken for a scale insect, is common in sugarcane trash used as mulch in the tropics.

(PHOTO BY THE AUTHOR)

SCHEFFLERA

Iron toxicity in seed-grown *S. arboricola:* Root rot early in the production cycle often leads to this disorder, but I have never seen it on arboricola cuttings.

(COURTESY OF JOHN GATTI)

SPATHIPHYLLUM

Myrothecium leaf spot of
Spathiphyllum: Spots are brown
and circular, with fruiting bodies
underneath.

(COURTESY OF ANN R. CHASE)

SPATHIPHYLLUM

Phytophthora blight of
Spathiphyllum is usually seen in
wet shadehouses, rarely in
greenhouses or interiorscapes.

(COURTESY OF MARSHALL HORSMAN)

SPATHIPHYLLUM

Calcium deficiency in
Spathiphyllum: The new leaves
fail to unfurl properly and often
have longitudinal "ribs."

(COURTESY OF JOHN GATTI)

SUCCULENT: CACTI

Thorn injury followed by secondary *Drechslera* infection in *Cereus peruv͐ ͐us:* This fungus is usually not systemic, except in cactus.

(COURTESY OF ED CLAY)

SYNGONIUM

Xanthomonas leaf spot of *Syngonium* White Butterfly: This bacterial spot is often accompanied by fungal infections.

(PHOTO BY THE AUTHOR)

SYNGONIUM

This *Syngonium* Infrared shows a combined infection of *Colletotrichum* and *Xanthomonas.*

(COURTESY OF OLAF RIBIERO)

YUCCA

Yucca cane going to the dump in Africa because of banana moth: Notice how the insect has mostly eaten the top and bottom parts of the cane.

(PHOTO BY THE AUTHOR)

YUCCA

Pythium root rot of yucca cane, aggravated by the excessive tropical rains: Highly compacted soil contributes to this problem.

(COURTESY OF ED CLAY)

OTHER FOLIAGE PLANTS:

AESCHYNANTHUS

Iron deficiency in aeschynanthus plant: Poorly aerated or wet soil is frequently the cause.

(COURTESY OF JOHN GATTI)

CARYOTA MITIS

Helminthosporium leaf spot of fishtail palm: Try to keep foliage dry and maintain good nutrient levels.

(COURTESY OF JOHN GATTI)

STRELITZIA

Iron deficiency in young white bird-of-paradise: High soil pH and poor soil aeration contribute to the chlorosis.

(PHOTO BY THE AUTHOR)

STRELITZIA

Xanthomonas blight in young white bird-of-paradise: This disease is rarely encountered in mature plants.

(COURTESY OF MARSHALL HORSMAN)

ZEBRINA

Leaf distortion of zebrina plant caused by thrips injury: The thrips feed on the unfurling leaves, which then grow out distorted.

(COURTESY OF JOHN GATTI)

other principal reason is that the grower can root and sell 3-inch peperomias from tip cuttings in three to five weeks.

The plant is very commonly seen in dish gardens, where its small size makes it compatible with crowded plantings. A few of the *P. obtusifolia* varieties are occasionally seen in small hanging baskets, typically 5 inches (12.5 cm).

VARIETIES

Some cultivars are upright plants, while others grow as sprawling vines. The naturally occurring P. obtusifolia has dark green, succulent, oval-shaped leaves. Minima is a dwarf version of the species, about half as big. P. obtusifola Variegata is an attractive plant with leaves that are creamy white in the outer portions, variegated dark and grayish green in the center. Albo-marginata is somewhat similar to Variegata, though the white margin is more narrow. Marble is a P. obtusifolia cultivar with a variegated pattern of green, white, and gray.

Other species of *Peperomia* are also common in the trade. Emerald-ripple, *P. caperata,* has very dark green, pleated leaves with reddish petioles. The watermelon peperomia, *P. argyreia,* has oval leaves with silver-gray stripes emanating from the center, and also reddish petioles. The red-edged peperomia, *P. clusiifolia,* is a larger plant, its oval foliage sporting a dark red edge. *P. griseoargentea,* the ivy-leaf peperomia, has more round foliage with sunken veins emanating from a rosette. *P. scandens,* the false philodendron, has succulent, heart-shaped foliage. This variety is good for hanging baskets.

PROPAGATION

Peperomias are easy to root from either stem cuttings, leaf cuttings, or leaf-and-eye cuttings. Watermelon, emerald-ripple, and ivy-leaf peperomias are usually propagated from leaf cuttings, with the rest usually rooted from stem cuttings. Most growers produce their own peperomias from in-house stock plants, though I have consulted for stock farms growing peperomias for cuttings in Guatemala, Honduras, and Costa Rica.

The quickest production is from stem tip cuttings, but leaf-and-eye cuttings are useful if stock is limited. The variegated cultivars should be propagated from stem cuttings which have one or more buds. Depending on the variety, stem cuttings are usually about 3 inches (7.5 cm) long, with two to four expanded leaves. Rooting hormone is not required. Two tips are generally rooted directly in 3-inch (7.5-cm) pots, or three to four tips in a 4-inch (10-cm) container. Water the freshly stuck cuttings in thoroughly but not heavily. Mist is generally not required, but don't let the medium dry out during the rooting process. A high percentage of cuttings should root in about two weeks.

CULTURE

One of the most important points for growing good peperomias is to use a potting mix with a low moisture-holding capacity. Peperomias do not like fluctuations between wet and dry conditions, so use a mix that won't hold a lot of water. Keep the peat percentage down, and use generous amounts of perlite, styrofoam beads, or bark in the mix. The plants should be grown in a covered greenhouse so that you have adequate control over soil moisture. Do your best to keep fairly consistent soil moisture levels for this plant.

Researchers say peperomias should be produced at light levels between 1,500 and 3,000 f.c. (16.1 and 32.3 klux). In hot areas you will do better on the darker end of that scale, and some varieties grow well at 1,200 f.c. (12.9 klux). Albo-marginata is one variety that does better under lower light. Temperatures are best maintained between 65 and 85F (18 and 29C). Cold damage can occur on most cultivars if the temperature drops below 50F (10C).

NUTRITION

Because peperomias are a fairly quick crop grown in small containers, liquid fertilizer is usually best. The crop can be too fast for slow-release, incorporated fertilizer, and top-dressing 3-inch (7.5-cm) containers is rather impractical. Dry fertilizers in this case also give consumers potential problems with soluble salts. When growing longer-term peperomias in larger pots, 2 grams of slow-release 19-6-12 or 18-6-8 per 4-inch (10-cm) pot are suggested, but only a three-month formulation.

Liquid fertilization about once a week with a 3-1-2 ratio, such as 9-3-6, is popular, using 150 ppm nitrogen. It is helpful to blend a conservative amount of a trace element preparation into the potting mix. Because of the fairly fast crop cycle, nutritional deficiencies are not common in production peperomias; they usually occur in the stock plants. Lack of nitrogen and potassium causes chlorosis in the older foliage. Lack of magnesium, a common occurrence in stock beds, causes a gradual reduction in color contrast in variegated varieties and marginal yellowing in older foliage of green varieties.

TABLE 20 Leaf analysis rating standards for *Peperomia obtusifolia*

Nutrient (%)	Very low	Low	Medium	High	Very high
Nitrogen	<2.25	2.25–2.89	2.90–4.50	4.51–5.50	>5.50
Sulfur	<0.20	0.20–0.24	0.25–0.75	0.76–1.25	>1.25
Phosphorus	<0.20	0.20–0.24	0.25–1.00	1.00–1.25	>1.25
Potassium	<2.50	2.50–3.99	4.00–6.50	6.51–7.50	>7.50
Magnesium	<0.25	0.25–0.39	0.40–1.20	1.21–1.50	>1.50
Calcium	<0.70	0.70–0.99	1.00–3.75	3.76–4.50	>4.50
Sodium			<0.20	0.21–0.50	>0.50
(ppm)					
Iron	<25	25–49	50–300	301–1000	>1000
Aluminum			<250	251–2000	>2000
Manganese	<25	25–49	50–300	301–1000	>1000
Boron	<20	20–24	25–50	51–100	>100
Copper	<4	4–6	7–100	101–500	>500
Zinc	<20	20–24	25–200	201–1000	>1000

Sources: Institute of Food and Agricultural Sciences, Apopka, Florida; Dr. Benjamin Wolf, Fort Lauderdale, Florida.

Notes: Common names include green peperomia, variegated peperomia. Sample of most recent fully mature leaves, no petioles.

DISEASES

Growers have to contend with several leaf spot diseases from time to time when growing peperomias. *Phyllosticta* causes tiny, black specks on the leaves, which occasionally coalesce into one-half-inch (1.25-cm) lesions with concentric light and dark rings. The Watermelon cultivar is especially susceptible. Sprays of Daconil (chlorothalonil), Dithane (mancozeb), or Chipco 26019 (iprodione) give good control. Moisture management is, of course, critical.

Cercospora causes tan to black leaf spots, which are raised on the underside, similar to edema. *Cercospora* is rather difficult for plant pathologists to isolate on *Peperomia,* and edema and *Cercospora* leaf spot are frequently confused. Edema is generally more widespread in a crop, whereas *Cercospora* symptoms are scattered. Sprays of a thiophanate methyl fungicide, such as Cleary's 3336 (thiophanate methyl) or Domain (iprodione), will help. Especially try to cover the undersides of the leaves. This may be better accomplished with a fogger or mist-blower type of sprayer.

Rhizoctonia causes large, mushy, dark brown to black leaf spots, especially on *P. obtusifolia* cultivars. The spots are irregular, with concentric light and dark rings. If you look closely, you may see tiny, reddish threads associated with the spots. Dump badly infected plants and spray the rest with Chipco 26019 (iprodione) or Domain (thiophanate methyl).

Southern blight, caused by the fungus *Sclerotium rolfsii*, causes plants to decline rapidly, especially in hot weather. White, threadlike mycelia and white to brown fruiting bodies are readily seen on plants and soil. Discard any obviously affected plants, then drench the remainder with either Terraclor (PCNB) or the insecticide Dursban (chlorpyrifos).

Pythium splendens and *Phytophthora parasitica* are the most common root rot diseases of peperomias. Plants rot at the soil line, and roots turn mushy and black. This especially happens when the medium is holding too much water. Dry the plants down and drench with Subdue (metalaxyl), Truban (etridiazole), or Aliette (fosetyl-aluminum).

A ring spot virus is occasionally a problem, especially on green *P. obtusifolia*. Yellow ghostlike spots appear, and foliage may be somewhat distorted. It is not a very serious problem, but dump virus-infected plants and try to avoid touching the remainder, as the virus is mechanically transmitted.

INSECT AND MITE PESTS

Numerous insects will attack *Peperomia*, but not frequently, fortunately. Caterpillars may cause chewing injury on occasion and leave large droppings. Sprays with Dipel *(Bacillus thuringiensis)*, Orthene (acephate), or Dursban (chlorpyrifos) control them. Mealybugs are found occasionally, especially on stock plants. The mealybugs tend to hide in the leaf axils and on leaf undersides. Sprays with Dycarb (bendiocarb) or Enstar (S-kinoprene) help, though repeated sprays are necessary. Scale insects, especially the aglaonema and proteus scales, are found occasionally. Treat as for mealybug. Thrips may cause scarring and leaf distortion, especially on younger foliage. Sprays of Avid (abamectin), Mavrik (tau-Fluvalinate), or Orthene are helpful.

Three different types of microscopic mites also attack peperomias: the broad mite, the cyclamen mite, and what should be called the peperomia mite, *Tarsonemus confusus*.

Since they are microscopic and cause similar damage, it is difficult to distinguish which mite is causing injury. Microscopic mites prefer to feed on the tender tissue of the developing shoot. The tip turns brown, and as the leaves grow out, they may be cupped, puckered, curled, or twisted, as the damage happens very early in their development. Sprays of Kelthane (dicofol), Pentac (dienoclor), or Avid (abamectin) are useful. When spraying for microscopic mites, use a wetting agent and switch miticides frequently.

Watch carefully for fungus gnat infestation in the rooting stage. The larvae feed on the bases of the cuttings and cause plant loss. Drench with Gnatrol *(Bacillus thuringiensis)*. Yellow sticky traps are also very useful in capturing the adults.

DISORDERS

Edema is a very common disorder of *Peperomia*. Edema is a physiological disorder wherein the plant takes in too much water, and the excess exudes through the intercellular spaces in the older leaves. This causes numerous tiny, darkish bumps on the undersides of the older foliage. The best control is to maintain a steady greenhouse environment and a consistent irrigation program. Because of their epiphytic nature, peperomias dislike low oxygen in the root zone. The plants take on a wilted appearance and grow very slowly. Try to dry the plants down and get some air back into the soil.

 Plants take up very little water at night.

Because of their fine roots, peperomias are quite sensitive to elevated soluble salts. The symptoms are very similar to those of moisture stress. Usually, by the time the grower notices it, the roots have already begun to die, and the plants start to defoliate. Fertilize peperomias conservatively and monitor your soluble salts. Water-stressed peperomias also tend to develop weak stems.

Avoid spraying with Cygon (dimethoate), Sevin (carbaryl), or Diazinon, as they can be phytotoxic on this plant.

TRICKS

Don't try growing peperomias in a shadehouse unless you live in a dry area with very consistent temperatures. A covered greenhouse with good climate control is preferable. Use a coarse potting mix, one that will tend to fall apart when you dump it out of a pot. Be sure the mix is at least relatively free of pathogens, and always use new pots and new medium.

Fertilizing with additional potassium increases variegation in the Albo marginata variety but not in *P. obtusifolia* Variegata. Growing stock plants under brighter light (60% shade) than production areas increases yield and produces stronger, more vigorous cuttings.

INTERIOR CARE

Try to maintain peperomias between 65 and 85F (18 and 29C). Temperatures below 50F (10C) injure the plants, so avoid cold or drafty areas. Peperomias, especially the less succulent types, like humidity. A minimum light level of 100 to 250 f.c. (1.1 to 2.7 klux) is desirable. Higher light than that is better, but don't go above 2,000 f.c. (21.6 klux).

Try to irrigate gently and consistently, maintaining steady soil moisture levels. Irrigate less during the winter and avoid extreme moisture fluctuations. Peperomias grow fairly slowly, so fertilize only three or four times per year with about one-half teaspoon per gallon (3.8 l) of 20-20-20.

REFERENCES

Blessington, T.M., and P.C. Collins. 1993. *Foliage Plants: Prolonging Quality,* 149-154. Batavia, Ill.: Ball Publishing.

Chase, A.R. 1988. *Phyllosticta* Leaf Spot of *Peperomia* Species and Cultivars in Florida. *Foliage Digest* (January): 4-5.

Henley, R.W., L.S. Osborne, and A.R. Chase. 1983. *Peperomias.* ARC-A Foliage Plant Research Note RH-1983-H. Apopka: Agricultural Research Center.

Seeley, J.G. 1983. Growing With Seeley. *GrowerTalks* (October): 14-16.

Watkins, J.V., and T.J. Sheehan. 1975. *Florida Landscape Plants,* 124-125. Gainesville: The University Presses of Florida.

PHILODENDRON

HABITAT

Philodendrons are among the most familiar foliage plants, having been in cultivation for centuries. The word *philodendron* means "tree loving" in Greek. About 200 species are found, mostly from tropical America. Three principal types of philodendrons exist. The vining or scandent types, such as *Philodendron scandens oxycardium,* the heart-leaf philodendron, are commonly grown as ornamental vines. The second group is of self-heading philodendrons. These plants, such as *P. wendlandii* and the hybrid Black Cardinal, grow upright on their own. The third class is of the erect-arborescent philodendrons. Plants in this group, best represented by *Philodendron selloum,* appear self-heading when young, but as they mature they become more woody and treelike.

Many philodendrons grow in South American rain forests as well as in the lowlands adjacent to the Andes. Since they grow naturally on and around trees, they prefer shaded, humid conditions. They are efficient users of moisture, and many have aerial roots to absorb water from humid air. Flooding can occur in their native habitats, so some philodendrons are ethylene sensitive.

USES

Because of the diversity of types, philodendrons are grown for a broad array of uses. Container sizes can range from 2 to 17 inches (5 to 42.5 cm). They're grown from single or multiple

Fig. 33 A 6-inch self-heading philodendron, from tissue culture. (Photo by the author)

217

cuttings in freestanding pots, while vining types are also frequently produced in hanging baskets or on totems. Philodendrons are useful as ground covers for interior and exterior landscapes. *Philodendron selloum,* also called the split-leaf or lacey-tree philodendron, is popular in USDA Zones 9 and 10 landscapes as a mass planting or a shrubbery border.

VARIETIES

The most widely produced philodendron is *P. scandens oxycardium,* the heart-leaf philodendron. It is frequently referred to incorrectly as *Philodendron cordatum,* which is actually a larger plant. Cuttings of heart-leaf philodendron were at one time smuggled onto the U.S. mainland from Puerto Rico on rum-runner boats. *P. panduriforme,* another vining type, is usually grown from air-layers for totems and dish gardens. It is a large, aggressive vine with heavily lobed foliage. It got to the U.S. foliage trade via a soldier who bribed the guard at an English botanic garden with a bottle of scotch in exchange for some cuttings.

P. selloum is grown as a freestanding specimen in many container sizes. Ultimately becoming woody and fairly erect, it features very deeply serrated foliage. Xanadu, a popular self-heading philodendron, looks somewhat like a miniature *P. selloum* with olive green foliage.

There is also a popular line of colorful hybrid philodendrons, many of which emanate from Bamboo Nurseries in Apopka, Florida. Black Cardinal is self-heading with dark maroon, almost black, succulent foliage. The new leaves are reddish. Prince of Orange, with orange new leaves, is a self-heading type that requires more light than many other philodendrons. Imperial Red and Imperial Green are also popular hybrid varieties. There are many other philodendrons, too many to list here.

PROPAGATION

Vining philodendrons are usually propagated from stem cuttings or leaf-and-eye cuttings. The heart-leaf philodendron is usually grown with multiple leaf-and-eye cuttings per pot or hanging basket, similar to pothos. The cuttings do not require any rooting hormone and very little in the way of mist. Tent propagation is also practiced. The larger vining types are also produced as air-layers, using sphagnum and aluminum foil. Most philodendrons root easily as long as they are given reasonable temperature and light levels. They also root easily in water.

P. selloum is generally grown from seeds, which are collected from tropical landscapes. Frequently, six to eight or more seeds are planted per cell in flats, with the seed barely covered with perlite. A minimum of 300 to 600 f.c. (3.2 to 6.5 klux) of light is required for germination. Most of the hybrid philodendrons are propagated from tissue culture and sold as liners.

CULTURE

Heart-leaf philodendrons should be grown between 1,500 and 3,000 f.c. (16.2 and 32.4 klux), with 2,000 f.c. (21.5 klux) being ideal. Stock plants should be maintained under somewhat brighter light. The minimum soil temperature should be 65F (18.3C), with a minimum air temperature of 75F (23.9C). It is best to avoid temperatures below 50F (10C), but the plants can tolerate 105F (40.5C).

P. selloum is usually grown much brighter, at 3,000 to 6,000 f.c. (32.4 to 64.8 klux). Many growers like to pot the liners into 6- or 10-inch (15- or 25-cm) containers and grow them in full sun until they reach maturity. The plants are then placed under shade for finishing, unless they will be sold for landscape use. *P. selloum* is also quite tolerant of cold. Depending on its age, it can tolerate temperatures close to freezing without injury, even surviving 22F (-6C) as a mature specimen.

Hybrid philodendrons are generally produced at between 1,000 and 2,000 f.c. (10.8 and 21.5 klux), though there are exceptions. Ideal daytime temperatures are 80 to 85F (26.7 to 29.4C), with night temperatures from 65 to 70F (18.3 to 21.1C). A hybrid philodendron should be grown on the dry side.

Philodendrons in general like a potting mix with a high moisture-holding capacity and good aeration. Mixtures of sphagnum or sedge peat, combined with bark, wood chips, perlite, or vermiculite are common. The target pH is 5.5 to 6.0. Philodendrons should be allowed to dry somewhat before irrigating.

NUTRITION

Most philodendrons are fairly heavy feeders, requiring generous amounts of dry or liquid fertilizers. A 3-1-2 ratio of N-P_2O_5-K_2O is generally used, as is extra magnesium. Smaller containers and baskets are often grown with 200 ppm nitrogen in constant liquid feed, derived from 24-8-16, liquid 9-3-6, or sometimes soluble 20-10-20. Leaches with clear water every few weeks are desirable. The larger philodendrons may be grown either with liquid feed or with granular, or coated, slow-release fertilizers. The larger philodendrons grow very well on a combination of dry and liquid fertilizer. Leaf size and

TABLE 21 Leaf analysis rating standards for *Philodendron panduriforme*

Nutrient (%)	Very low	Low	Medium	High	Very high
Nitrogen	<2.25	2.25–2.49	2.50–4.50	4.51–6.00	>6.00
Sulfur	<0.15	0.15–0.19	0.20–0.50	0.51–1.00	>1.00
Phosphorus	<0.18	0.18–0.24	0.25–0.50	0.51–1.00	>1.00
Potassium	<1.50	1.50–1.99	2.00–3.75	3.76–5.00	>5.00
Magnesium	<0.20	0.20–0.24	0.25–0.50	0.51–1.00	>1.00
Calcium	<0.70	0.70–0.99	1.00–2.50	2.51–3.50	>3.50
Sodium			<0.21	0.21–0.50	>0.50
(ppm)					
Iron	<50	50–59	60–200	201–1000	>1000
Aluminum			<251	251–2000	>2000
Manganese	<30	30–39	40–200	201–1000	>1000
Boron	<15	15–19	20–50	51–75	>75
Copper	<5	5–7	8–100	101–500	>500
Zinc	<20	20–25	26–100	101–1000	>1000

Sources: Institute of Food and Agricultural Sciences, Apopka, Florida; Dr. Benjamin Wolf, Fort Lauderdale, Florida.
Note: Sample of most recent fully mature leaves, no petioles.

growth rate are good indicators of fertility status in *Philodendron*. Leaf size and internode length in most varieties quickly decrease if feed rates drop below optimum levels.

Philodendrons usually don't have much in the way of trace element problems. Many show marginal chlorosis of the older foliage when deficient in magnesium. This especially happens in older plants or plants that are cut back several times. It helps to keep magnesium in the spray program on a regular basis. Lack of calcium causes root tips to die, followed by chlorosis, leaf distortion, and shoot tip death. Sprays of a combination calcium and magnesium chelate result in good, strong philodendrons with large leaves. Iron deficiency symptoms in hybrid philodendrons usually indicate root disease.

Plants need most of their phosphorus early in their development.

DISEASES

Bacterial pathogens can be a particular problem on *Philodendron* at times. Two types of *Erwinia* are common, *E. chrysanthemi* and *E. carotovora*. Most cultivated philodendron varieties are susceptible. Plants develop water-soaked lesions, and the rapid decomposition of the tissue results in a foul odor. Symptoms can develop within two days under the right conditions. Adequate spacing, good nutrition, and drying the plants out are helpful cultural controls. Sprays of Agrimycin 17 (streptomycin sulfate), tank mixes of Dithane (mancozeb) plus a copper fungicide, or Phyton 27 (picro cupric ammonium formate) are somewhat helpful. Sanitation and cultural controls are best. *Pseudomonas cichorii* causes a bacterial blight similar to *Erwinia*, though it is less wet and mushy. Treat with a similar combination of cultural and chemical controls.

Xanthomonas campestris pv. *dieffenbachiae* is a fairly serious disease of *Philodendron*, especially on the heart-leaf philodendron. It is commonly called red-edge. The older leaves have reddish margins, and the infection may spread to internal portions of the leaf. Control measures are similar to those for *Erwinia*, and infected stock should usually be discarded.

Common fungal blights include *Phytophthora parasitica*, which causes dark brown, water-soaked, irregularly shaped lesions. The leaf spots are typically one-half to 1 inch (1.25 to 2.5 cm) wide, and they tend to happen in wet conditions. Dry the plants out, increase air movement, and either spray with Dithane (mancozeb) or Daconil (chlorothalonil) or drench with Subdue (metalaxyl). *Dactylaria* causes numerous pinpoint, tan leaf spots, frequently on younger foliage. In addition to cultural controls, sprays with Dithane, Daconil, or one of the thiophanate methyl fungicides are helpful. *Sclerotium rolfsii*, commonly called southern blight, may attack many philodendron cultivars in hot weather. White fungal mycelia are very visible on the basal plant tissue as well as on the soil. Cool the greenhouse down, if possible, and either drench with Terraclor (PCNB) or spray with the insecticide Dursban (chlorpyrifos). *Rhizoctonia* can cause a sudden blight of *Philodendron* in summer. Large, brown lesions will form very quickly, frequently near the center of the plant. Dry it down and spray with either Daconil or Chipco 26019 (iprodione).

Dasheen mosaic virus is not very common on *Philodendron*, but it does attack a number of different species. Mosaic symptoms within the leaf, stunting, and leaf distortion are apparent. The virus is spread by both aphids and cutting tools. There are no controls, so it is best to discard infected plants.

Stock beds are sometimes infected with burrowing nematodes. The root systems are greatly reduced, so the plants lack vigor and fail to respond to good light and fertilizer levels. Infected soil and plants should be either steamed or fumigated to kill the nematodes so they don't contaminate other areas or landfills, then dumped.

INSECT AND MITE PESTS

Aphid attacks are common with new growth, especially in spring or when growth is rapid. Yellow, green, and brown aphids cause leaf cupping and distortion. Insecticidal soaps are useful, as are sprays of Orthene (acephate), Mavrik (tau-Fluvalinate), and numerous others. Caterpillars, including the omnivorous leaf roller, may chew holes in succulent philodendron foliage. Exclude moths and butterflies from the greenhouse, if possible, and spray with Dipel or one of the other *Bacillus thuringiensis* insecticides if caterpillars are small. Numerous other insecticides can be used, including Sevin (carbaryl) or Dursban (chlorpyrifos).

Philodendron vines make great hiding places for mealybugs, which can be extremely troublesome. Their white, cottony masses and accompanying sooty mold can be found on many parts of philodendron vines. Sprays of Dycarb (bendiocarb) or Mavrik (tau-Fluvalinate) control them if you have good coverage and use repeated sprays. Root mealybugs also attack a number of philodendrons. They look like regular mealybugs, only somewhat smaller and rounder. A soil drench with Diazinon usually controls them.

Several kinds of scale insects attack philodendrons, and they are occasionally found on leaves, petioles, or stems. Treatment is similar to mealybug, though granular Di-syston (disulfoton) can be very helpful. Finally, thrips may cause distortion or scarring of foliage. They frequently affect one side of the leaf more than the other, as their feeding often occurs in the tube of the unfurling leaf. Sprays of Mavrik (tau-Fluvalinate) or Orthene (acephate) are helpful.

DISORDERS

Low winter soil temperatures cause philodendrons to grow slowly, with thin stems and short internodes. The bird's-nest fungi, *Crucibulum* and *Sphaerobolus,* appear as small, brown to black disks on the undersides of philodendron leaves; these are not scale insects. Bird's-nest fungus frequently propagates itself on wood products in the medium or in mulch on greenhouse floors. Captan sprays have been successful, though Captan is not registered for use on *Philodendron* in the United States.

Philodendron leaves may become too large if excessive nitrogen is applied. Small foliage frequently indicates inadequate nitrogen. If both the leaves and the stems of philodendrons are small and weak and the rooting performance is poor, lack of light or fertilizer is generally the cause.

Varieties such as Xanadu or Pluto may turn distinctly reddish when exposed to chilling temperatures. Black Cardinal routinely develops round, orange to tan spots on the older foliage. This is actually spider mite injury, frequently caused by European red mite. Leaf distortion in Black Cardinal may be caused by thrips injury or boron deficiency. *Philodendron selloum* may develop sunburn on the older leaves even if it is well acclimated to full sun. The problem happens either after plants are spaced or when they are not given sufficient irrigation.

TRICKS

Increasing nitrogen fertilization decreases incidence of red-edge disease in the heart-leaf philodendron. This variety also prefers warm soil temperatures for best growth and development. *P. selloum* can be kept relatively free of *Erwinia* during rainy summers if sprayed twice a week, alternating Agrimycin 17 with copper plus mancozeb. The disease is much less severe on mature foliage. When trying to control disease in philodendrons, it helps a great deal to irrigate when the drying conditions are good.

Foliar sprays of calcium help philodendrons develop good, thick leaves. Add extra magnesium to the liquid or dry feed when maintaining stock plants. Don't use much vermiculite in the potting medium if growing in a shadehouse in a rainy environment, as the mix may hold too much moisture. Vermiculite is fine for philodendrons in greenhouses. Drip irrigation is useful for controlling disease on larger philodendrons and totems. When mixing Dithane (mancozeb) and a copper fungicide for bacterial disease control, it helps to let the materials mix in the spray tank for 60 to 90 minutes prior to spraying.

INTERIOR CARE

Philodendrons are best maintained with minimum light levels of 75 to 150 f.c. (0.8 to 1.6 klux). Heart-leaf philodendron tolerates 50 f.c. (0.5 klux) somewhat, while *P. selloum* requires more light. Temperatures between 65 and 85F (18 and 29C) are best. Avoid cold drafts, and try not to let the temperature drop below 50F (10C) for any period of time. Fertilize lightly two to three times per year, more often if you have very good light levels.

Philodendrons tend to prefer humid interior environments. If that is difficult, setting the plants on a tray filled with moistened pebbles helps maintain a humid environment around the plants. Aerial roots may be removed if they are unsightly because they serve little function in a low-humidity interior.

REFERENCES

Blessington, T.M., and P.C. Collins. 1993. *Foliage Plants: Prolonging Quality,* 155-160. Batavia, Ill.: Ball Publishing.

Chase, A.R. 1993. Common Diseases of Philodendron. *Southern Nursery Digest* (May): 20-21.

Conover, C.A., L.S. Osborne, and A.R. Chase. 1984. Heart-Leaf Philodendron. *Florida Nurseryman* (May): 15-17, 42.

Harkness, R.W., and R.B. Marlatt. 1970. Effect of Nitrogen, Phosphorus, and Potassium on Growth and *Xanthomonas* Disease of *Philodendron oxycardium. J. Amer. Soc. Hort. Sci.* 95 (1): 37-41.

Henley, R.W., L.S. Osborne, and A.R Chase. 1985. *Lacy-Tree Philodendron.* AREC-A Foliage Plant Research Note RH-85-E: 1-3. Apopka: Agricultural Research and Education Center.

Hershey, D.R., and R.H. Merritt. 1987. Calcium Deficiency Symptoms of Heartleaf Philodendron. *Hortscience* 22 (2): 311.

Miller, H. 1984. Variety Showcase. *Greenhouse Grower* (November): 84-85.

Norris, C.A. 1987. The Business of Breeding at Bamboo Nurseries. *GrowerTalks* (January): 60-63.

Poole, R.T. 1992. Effects of Medium Temperatures and Magnesium Application Rate on Growth of *Philodendron scandens oxycardium. Foliage Digest* (March): 7-8.

Schmidt, P. 1979. Philodendrons. *House Plants and Porch Gardens* (September): 52-60.

PHOENIX
Date Palm

HABITAT

Phoenix was the ancient Greek name for the date palm. There are about 17 species of date palms, which are generally native to Africa and Asia. The two principal date palms in the foliage trade are actually quite different. *Phoenix roebelenii,* the pygmy date palm, is produced in large numbers by growers in several regions of the world. The species was named after a Mr. Roebeleni. *P. roebelenii* generally does not grow larger than 6 feet (1.8 m), although I have occasionally seen ones 10 feet (3 m) tall. Like most date palms, it has spiny petioles. Pygmy date palms are native to Laos, but they are also found from northeast India to Burma. They frequently grow along riverbanks, sometimes under the shade of other trees, and sometimes out in the open. In culture pygmy dates do well in moist conditions and are quite adaptable regarding light levels.

The other phoenix palm that is primary in the trade is *Phoenix reclinata,* the Senegal date or wild date palm. It is a somewhat variable species widely distributed through-out tropical and southern Africa. *P. reclinata* typically grows about 20 feet (6.1 m) tall, though in Africa it has been known to reach 50 feet

Fig. 34 A three-year-old *Phoenix roebelenii,* grown in Homestead, Florida. (Courtesy of Ed Clay)

225

(15 m). The leaves can be as long as 9 feet (2.7 m). *P. reclinata,* a clustering palm, has trunks that almost always lean at an angle, hence the species name. Like *P. roebelenii,* it is often found along streambanks in its native habitat, though sometimes it is seen in coastal or desert locations.

USES

Pygmy dates—durable, attractive, dwarf palms—are considerably tougher than many other palms used for interiors. Growers usually produce them in containers ranging from 6 to 17 inches (15 to 42.5 cm). Two or three seedlings are sometimes placed in larger containers for marketing as a multiple-trunked palm. They are popular in planters, urns, and large tubs. Mature specimens are also used in planter beds in interiors, where they are quite durable.

Senegal date palms are quite large, so their interior use is limited primarily to shopping malls and large atriums. They are also seen along medians and near entrances to larger buildings. Because it frequently has 10 to 20 stems, *P. reclinata* is often placed as a focal point in a circular planter bed.

VARIETIES

Only the species varieties of these two palms exist, though *P. reclinata* is somewhat variable genetically in size, number of trunks, and other small details. Phoenix palms tend to cross-pollinate and hybridize with each other naturally. You will therefore occasionally see *P. roebelenii* hybrids crossed with *P. reclinata* or other phoenix palms. The hybrids usually have characteristics intermediate between those of the parents.

PROPAGATION

Palms are almost always propagated from seed. Phoenix palms have a very curious characteristic: the seedling actually develops a short distance away from the actual seed. The first shoot to emerge from a phoenix seed, the cotyledonary petiole, grows down into the soil, occasionally fairly deeply, where it then begins to swell. The radicle, or first root, and shoot emerge from this swelling. Such germination is referred to as remote germination.

Phoenix palm seeds are generally mass-planted in plastic flats or on raised beds. A typical, well-drained potting soil is generally used. Cover the seeds with soil but don't plant them deep. A seedling can be transplanted to a larger container once its first true leaf emerges and preferably before its roots become intertwined with those of other seedlings. *P. roebelenii* seed can be stored up to eight months in a plastic bag after dusting with a little Thiram (thylate). Storage temperature should be 65 to 75F (18.3 to 23.9C).

CULTURE

Pygmy dates are generally grown in mixtures of peat, bark, and sand. Either sphagnum or sedge peat, or a mixture of the two, can be used. I like to lime the mix to pH 6.5 or so with extra dolomite, as phoenix palms have high magnesium requirements. Light levels can range from full sun to 73% shade, though 50 to 63% is probably ideal. Because of its size, *P. reclinata* is frequently field grown in well-drained, sandy sites with irrigation. Some nurseries produce *P. reclinata* multiple-trunked in containers, as well. Both palms are fairly adaptable because they tolerate variable moisture regimes.

Phoenix palms can sustain high temperatures and can take surprisingly cold weather, too. In mature specimens first cold injury has been reported at 19.4F (-7C). *P. roebelenii* suffers severe injury around 17.6F (-8C). *P. reclinata* is a little tougher, suffering severe damage around 13.1F (-10.5C). Interestingly, when Hurricane Andrew hit Homestead, Florida, a major foliage-producing area, entire neighborhoods were blown away, but pygmy date palms suffered very little injury.

NUTRITION

Phoenix palms in general have fairly particular nutrient requirements. They need more potassium, magnesium, and manganese than most plants. Young plants are frequently fertilized with liquid fertilizer containing 150 ppm nitrogen and potash, frequently from 20-10-20. More mature specimens can receive liquid fertilizer at 300 ppm N and K_2O, preferably with supplemental magnesium. Granular fertilizers with 3-1-2 ratios are commonly used, but you will do better with a 3-1-3 mixture, such as 12-4-12, with at least 3% magnesium and 1% manganese. It helps a great deal to have some of the potassium derived from coated or slow-release forms.

Magnesium deficiency (see "Disorders") can take up to nine months to correct in phoenix palms. The deficiency is therefore best prevented with extra magnesium in the fertilizer program. It is also useful to spray periodically with manganese sulfate at one-half pound per 100 gallons (225 g per 400 l) to help prevent manganese deficiency.

TABLE 22 Leaf analysis rating standards for *P. roebelenii*

Nutrient (%)	Very low	Low	Medium	High	Very high
Nitrogen	<1.6	1.6–1.9	2.0–3.0	3.1–3.8	>3.8
Sulfur	<0.15	0.15–0.20	0.21–0.75	0.76–1.25	>1.25
Phosphorus	<0.11	0.11–0.14	0.15–0.75	0.76–1.0	>1.0
Potassium	<1.01	1.01–1.24	1.25–2.75	2.80–3.50	>3.50
Magnesium	<0.22	0.22–0.29	0.30–0.75	0.76–1.00	>1.00
Calcium	<0.60	0.60–0.74	0.75–2.00	2.01–3.00	>3.00
Sodium			<0.11	0.11–0.25	>0.26
(ppm)					
Iron	<40	40–49	50–300	301–1000	>1000
Aluminum			<251	251–1000	>1000
Manganese	<42	42–49	50–250	251–1000	>1000
Boron	<12	12–14	15–60	61–100	>100
Copper	<4	4–5	6–50	51–500	>500
Zinc	<18	18–24	25–200	201–1000	>1000

Sources: Institute of Food and Agricultural Sciences, Apopka, Florida; Dr. Benjamin Wolf, Fort Lauderdale, Florida.

Notes: Common names include pygmy date palm, Canary Island date palm. Sample of most recent fully mature leaves, no petioles.

DISEASES

Phoenix palms produced for the foliage trade are not particularly disease prone. Brown fungal leaf spots, usually caused by *Helminthosporium* or *Gloeosporium,* occasionally occur. Sprays of Dithane (mancozeb) or Chipco 26019 (iprodione) usually control them. Keeping foliage dry is, of course, helpful.

Phoenix palms have fairly strong root systems, but overwatering can encourage root rot from *Pythium, Fusarium,* or *Rhizoctonia* to occur. A decent soil mix and letting the medium dry out between irrigations is usually sufficient. If root rot occurs, growers usually drench with either Chipco 26019 (iprodione) or a thiophanate methyl fungicide, frequently in combination with Subdue (metalaxyl).

The bacterium *Erwinia* occasionally causes bud rot in pygmy dates, usually following flooding or severe cold injury. Chemical treatments are generally not helpful in such a case.

Phoenix reclinata does have a few severe disease problems, though they are rarely encountered in foliage nurseries. *Ganoderma* may at times cause a root and trunk rot of *P. reclinata*. The disease is characterized by tan sporophores or conchs, usually found at the base of the trunk on the north side. There is no chemical control, and the disease is fatal. Lethal yellowing, caused by a mycoplasma-like organism, causes flowers and fruit to drop. The palm will yellow and die back, usually over a period of three months or so. *Fusarium* wilt can also attack *P. reclinata*, causing frond dieback and ultimately death. There are no chemical controls for either lethal yellowing or *Fusarium* wilt currently, but fortunately, these diseases are not frequently encountered in foliage production.

We have a collection of Ganoderma conchs in the lab which we call the Conch Republic.

INSECT AND MITE PESTS

Phoenix palms in general have fairly tough leaves, so insect pests are usually only a minor problem. Scale insects may attack both Senegal dates and pygmy dates, Florida red scale and the oleander scale are two of the most common. They are more numerous on older leaves, generally, where they proliferate and cause gradual decline of the palms; sprays of Cygon (dimethoate) or Supracide (methidathion) are helpful.

Spider mites occasionally attack *P. roebelenii*. Predator mites *(Phytoseiulus)*, insecticidal soaps, or miticides, such as Pentac (dienoclor), usually control them. Palm weevils may attack *P. reclinata*, causing potentially fatal bud injury. Sevin (carbaryl) has been somewhat successful in controlling them.

DISORDERS

Use of the preemergent herbicide Ronstar (oxadiazon) can cause small, round, orange leaf spots on the foliage of *P. roebelenii*. The spots look somewhat like a disease. A curious ladder disorder sometimes develops on *P. roebelenii*, where the tips of the newer leaves remain stuck together, forming a ladderlike symptom. Deficiency of chloride, an essential nutrient, is usually the cause.

Lack of nitrogen in phoenix palms causes light green foliage and reduced growth. Potassium deficiency, a common occurrence, causes yellowing or spotting in the older leaves, frequently with some necrotic spotting. The leaflet tips also frequently die back. Magnesium deficiency causes a golden or yellow pattern in the older leaves, while new foliage remains green. Drenches or topdressings with magnesium sulfate help cure the problem, though the treatment should be repeated at least once to ensure full response. Badly yellowed leaves will likely not recover, however.

Manganese deficiency is especially common in pygmy date palms. New foliage appears withered and somewhat lime green, frequently with brown streaks in the withered foliage. This especially happens in alkaline soils and with cold temperatures. Iron deficiency causes emerging leaves of phoenix palms to be bright yellow. More frequent in *P. roebelenii* than *P. reclinata,* this will happen when soil aeration is poor or when palms are planted too deeply. Sulfur deficiency in phoenix palms looks much like iron deficiency, though it is somewhat less common.

TRICKS

Phoenix palm foliage repels water quite readily. Unless the label indicates otherwise, always use a wetting agent or surfactant when spraying date palms. Phoenix palms are not very efficient in absorbing nitrogen and potassium into their foliage. Trace element absorption seems reasonably good, however, and while it hasn't been demonstrated experimentally in palms, many feel that adding potassium nitrate to a trace element spray will aid in absorption. Top-dressing with granular Sul-po-mag (0-0-22-11) helps prevent or cure magnesium and potassium deficiencies. If you need to repeat the application, do it every three months.

Also, a few tablets of water softener salt applied as a topdress will act as a slow-release source of chloride.

Phoenix palm seeds can be tested for viability by cutting the seeds in half and placing the half that contains the embryo in a 1% solution of Tetrazolium in the dark. Viable embryos turn red or pink, but nonviable embryos are not likely to change color. Plant seeds at uniform depth to have more uniform seedling development.

INTERIOR CARE

Phoenix palms do better with as much light as you can give them in the interior environment. The minimum is probably around 300 f.c. (3.2 klux). New growth becomes stretched and weak over time if light is inadequate. Do not fertilize unless you have adequate light for reasonably normal growth. When fertilizing, use no more than a teaspoon of soluble 20-20-20 per gallon (3.8 l), preferably with one-fourth to one-half teaspoon of magnesium sulfate.

Inspect the plants occasionally for scale, mealybug, and mite activity. These palms tolerate being pot bound, though in an old specimen numerous surface roots will become visible and can eventually break the container.

REFERENCES

Bailey, L.H., and E.Z. Bailey. 1976. *Hortus Third*, 862-863. New York: Macmillan.

Broschat, T.K. 1992. Effect of Manganese Source on Manganese Uptake by Pygmy Date Palms. *Tropicline* 5 (1): 1-4.

Broschat, T.K., and A.W. Meerow. 1990. Palm Nutritional Disorders in the Landscape. *The Green Keeper* (November 30): 2.

Henley, R.W. 1984. Cold Injury of Palms. *Foliage News* 9 (10): 1-3.

Meerow, A.W. 1991. Palm Seed Germination, Part 1. *Tropicline* 4 (1): 1-3.

Simone, G.W., and G. Cashion. 1996. Fusarium Wilt of Canary Island Date Palms in Florida. *Landscape and Nursery Digest* (May): 28-31.

Stewart, L. 1993. *A Guide to Palms and Cycads of the World*, 152. Sydney, Australia: Augus & Robertson.

PIGGYBACK PLANT

(See Other Foliage: Tolmiea*)*

PRAYER PLANT

(See Maranta*)*

RAVENEA
Majesty Palm

HABITAT

Ravenea palms are new to the foliage trade, having been commercially produced, at least in Florida, only since about 1990. This palm was introduced to the foliage trade by my friend Dr. Henry Donselman, a California palm consultant. Very little information has been published about this palm, so much of this comes from my personal experience.

Ten species exist, eight from Madagascar and two from the nearby Comoro Islands. The one cultivated species is *Ravenea rivularis,* the majesty palm. The majesty, a stout, single-trunked, pinnate palm, somewhat resembles the kentia palm when young and the royal palm *(Roystonea)* when mature. In its native habitat the majesty grows near riverbanks, hence the species name *rivularis.* It is found in rain forests and swampy areas in subtropical to warm temperate zones. Thus, it likes moist conditions and rich, fertile, humic soils. It is fast growing and rather cold tolerant.

A friend of mine once called the majesty palm the palm of the future, due to its attractive appearance, its thick trunk, and its rapid growth. Larger pot sizes are generally produced, the most common being 10, 14, and 17 inches (25, 35, and 42.5 cm). I suspect that within a few years

Fig. 35 A 24-inch *Ravenea rivularis* (majesty palm) at the author's home in West Palm Beach, Florida. (Photo by the author)

232

we will see mature majesty palms in shopping malls, atriums, and large banks, once plants in cultivation reach appropriate size. For now, raveneas are used as freestanding, single-trunk specimens in large planters and as containerized accent plants in room corners. Majesty palms will also likely become popular landscape palms for USDA Zone 9.

Varieties

Only the species *Ravenea rivularis* is in commercial production in the foliage trade to date. None of the other raveneas are yet grown commercially, and there are no named varieties.

Propagation

Majesty palms are propagated solely from seed. The fruits are red when mature, and the pulp is cleaned from the seed prior to planting. Majesty palm seed loses viability quickly, so it should be planted promptly. Barely cover seeds with a well-drained potting soil, either in seed flats or in deep cell trays. Fresh seeds should germinate quickly. The seed beds can be maintained in either full sun or shade. Once the seedlings are up, most growers fairly quickly transfer them directly to 6- or 10-inch (15- or 25-cm) pots.

Culture

So far, majesty palms are produced only as single-trunk specimens. They require well-drained, fertile, rich soil with good moisture retention. Potting mixes usually contain at least 50% peat and 10% coarse sand, with the balance of the mix made up of bark or wood chips. The palms are frequently started in full sun and moved to 63 or 73% shade as they approach maturity. Majesty palms may blow over when grown in full sun and exposed to wind, however, so it helps to anchor the outer two rows of plants in the block with some sort of staking.

Leaf color tends to be better in shade, though when fertilized correctly they can be quite dark green, even in full sun. In shade or greenhouses, production light levels are usually 3,000 to 4,500 f.c. (32.3 to 48.4 klux). As long as the soil is moist, majesty palms tolerate temperatures up to 100F (37.8C). The palms are tolerant of cold but will be damaged around 24F (-4.4C). Cold-damaged raveneas recover fairly quickly from cold injury, however.

Nickel is an essential plant nutrient.

NUTRITION

Nutrient requirements of majesty palms are significantly higher than those of other palms and higher than most foliage plants in general. Growers frequently use granular fertilizers with approximately a 3-1-2 ratio of N-P_2O_5-K_2O, but at about 50% greater than the normal dosage. When using slow-release, incorporated fertilizers, such as 18-6-8 or 19-6-12, I have seen good growth at double the recommended rate. It is possible to injure majesty palms from high soluble salts, but it is not easy if soil moisture is good. I generally recommend coming back and refertilizing when soil conductivity readings reach 1.0 mmhos by the 2-to-1 dilution method. Constant liquid feed at 300 to 350 ppm nitrogen from soluble 24-8-16 or liquid 9-3-6 is also used.

I strongly suspect that raveneas are acid-loving plants, though this has not been formally proven to date. Acid-loving plants are really iron-loving plants, and raveneas certainly have very high iron requirements. They also need more magnesium and manganese than most plants. Incorporating trace elements into the potting mix is essential, as are trace elements in the fertilizer program. Many growers also like to supplement with foliar sprays of magnesium products and trace element preparations.

DISEASES

A number of diseases have already been reported on majesty palms, though none is particularly severe. The *Helminthosporium* complex, caused by *Exerohilum* and several related fungi, causes numerous circular, medium to dark brown leaf spots about one-eighth inch (3.1 mm) or more across. Controls include keeping foliage dry and spraying with Dithane (mancozeb) or Chipco 26019 (iprodione). *Colletotrichum* causes anthracnose, which is a somewhat larger and more oval brown leaf spot. Sprays of Dithane or a thiophanate methyl fungicide, such as Cleary's 3336 or Domain, are useful.

Several root rot diseases have been reported in *Ravenea,* though, again, none is particularly serious. *Fusarium, Rhizoctonia,* and *Pythium* may all cause some degree of root rot at times, especially when drainage is poor or soil moisture is constantly on the high side. *Fusarium* and *Rhizoctonia* can be controlled to some degree with drenches of a thiophanate methyl fungicide or Chipco 26019 (iprodione). In addition to drying the soil down, control *Pythium* with either Subdue (metalaxyl), Aliette (fosetyl-aluminum), Truban (etridiazole), or Banol (propamocarb).

In California majesty palms are known to be very susceptible to pink rot, which is caused by the fungus *Gliocladium vermoeseni*. This fungus does not like higher temperatures, which probably explains why pink rot is relatively rare on majesty palms in Florida. Sprays of a thiophanate methyl fungicide or Dithane (mancozeb) may be helpful, but the disease can be difficult to control. Wounding facilitates infection.

INSECT AND MITE PESTS

Like with many other palms, two-spotted mites occasionally cause problems with majesty palms. Growers usually use either a predator mite program with *Phytoseiulus* or spray periodically with one of the various foliage miticides, including Avid (abamectin), Pentac (dienoclor), or insecticidal soaps. I have observed Florida red scale and oleander scale on majesty palms on a few occasions. Granular top-dressing with Di-syston (disulfoton) controls scales, as do such sprays of Supracide (methidathion).

DISORDERS

It is difficult to keep the new majesty palm foliage fully green when the trees are growing rapidly. Growers frequently keep the pH too high for this palm, and with high fertility and warm conditions, the iron demands of the plant are not met. Lack of iron causes the typical chlorosis in the first one or two emerging leaves. The leaves after that tend to remain fully green, though in a severe case the whole palm may turn yellow. Lack of manganese can cause frizzle-top, characterized by a distortion of the emerging frond. This disorder is not particularly common in container production, though it will occur in landscapes with alkaline soils. Magnesium deficiency is characterized by broad areas of golden yellow to orange chlorosis in the older leaves.

TRICKS

As I indicated earlier, I strongly suspect majesty palms to be acid-loving plants. I think the grower would do perfectly well growing raveneas at a soil pH of 5.0, with abundant trace elements and fertilizers incorporated into the mix. If chlorosis develops, rapid response can be achieved by drenching with an EDDHA iron chelate, typically at the highest rate. To help minimize transfer of disease, try to trim majesty palm fronds only in hot weather, preferably at least 86 to 92F (30 to 33.3C). Pink rot should not be active at those temperatures.

Majesties can be grown with less labor by planting the seeds in individual deep cells. Once they sprout, transplant directly into a 10-inch (25-cm) container, but do not plant majesties deeply. Make sure the flare at the base of the trunk is not covered with soil. Also, the broad, somewhat stiff fronds catch a lot of wind, so adding at least 10 to 15% coarse sand by volume to the medium helps keep the plant from blowing over.

INTERIOR CARE

One reference indicates that majesty palms may not be particularly good interior plants. Neither are areca palms, though hundreds of thousands are sold annually. Information at this point is sketchy, but I would try to maintain majesty palms indoors with as much light as I could give them. If they must be maintained fairly dark, it helps to move them to a brighter location, such as a covered porch or a shaded greenhouse, for a couple of weeks to revitalize them. Do not move them back out into full sun, though, or foliar burn will occur.

Water interior majesties regularly but don't overwater. If they get dry, they tend to lose older leaves one at a time. Do give them a little fertilizer here and there, but if light levels are low, new growth may be very weak and stretched. If that is the case, then refrain from fertilizing. Put Epsom salts in the irrigation water from time to time to help the plant maintain adequate magnesium.

REFERENCES

Chase, A.R., and T.K. Broschat. 1991. *Diseases and Disorders of Ornamental Palms,* 25. St. Paul, Minn.: The American Phytopathological Society.

Donselman, H. March 1996. Telephone conversation with the author.

Stewart, L. 1994. *A Guide to Palms and Cycads of the World,* 168. Sydney, Australia: Augus & Robertson.

RHAPIS

Lady Palm

HABITAT

Rhapis, commonly called the lady palm, has been in cultivation since the 1600s. Of the 12 species, the main cultivated one comes from southern China. The word *rhapis* is Greek for "needle." All lady palms are short, dense, clustering palms. Their persistent leaf bases break down into a web of fibers on the trunk. They grow naturally as understory plants in dense subtropical evergreen forests. They therefore prefer fairly heavy shade, though they tolerate high light. With their variable native climate, rhapis palms tolerate temperature fluctuations very well. Though they seem to favor humid conditions, they adapt to low humidity, as well.

Fig. 36 *Rhapis excelsa*, showing iron deficiency in the newer leaves. (Photo by the author)

USES

Rhapis excelsa has been cultivated since the mid-1600s and was described by the botanist Aiton in 1789. *Rhapis humilis* was collected from China by the Japanese around 1700. The palm was brought to the United States in the early 1900s, where the original specimens can still be found growing at the Huntington Botanical Gardens in California.

Larger *Rhapis* specimens are commonly produced in 10-, 14-, and 17-inch (25-, 35-, and 42.5-cm) containers. They are used as freestanding container plants for offices and commercial buildings. Large lady palms are often installed in planter beds in malls and atriums, where they help create a somewhat tropical, oriental effect. An interesting look is achieved with small rhapis palms in hanging baskets on mezzanines.

In the Orient itself, larger rhapis palms are produced in ornate ceramic containers. Miniature *Rhapis* varieties are also popular in the Far East in small, bonsai-type pots for tables and desks.

VARIETIES

Rhapis excelsa is the principal rhapis palm grown in Florida. The Japanese have come up with a number of dwarf *R. excelsa* varieties, which are usually propagated by division, including Daruma, Koban, Tenzan, Kodaruma, and the variegated Zuikonishiki.

Rhapis humilis is more commonly grown in cooler climates, such as California. Nomenclature of rhapis palms has been confusing and inconsistent at times. *R. humilis* is commonly called the dwarf lady palm or slender lady palm. This is interesting, as *R. humilus* ultimately becomes a larger plant than *R. excelsa*, though the trunks clearly are more slender. The variety Thailand, also called Thai Dwarf, is a selection of *Rhapis subtilis*.

PROPAGATION

Rhapis excelsa may be propagated either by seed or by division. The small, black seeds germinate in 50 to 60 days if they are fresh. Some growers plant them directly in 6-inch (15-cm) pots, while others like to plant in seed beds and then transplant. You usually need to have at least six seedlings per pot, and you can count on suckers from the larger varieties. Some growers plant two or three 6-inch liners into a larger container for fullness. *R. excelsa* may be produced as a field-grown palm in full sun. The clumps are periodically dug and transplanted into large containers in shadehouses, where they require about four months to acclimate.

Rhapis humilis rarely produces seed and is therefore generally propagated by division. The dwarf Japanese *R. excelsa* varieties tend not to come true from seed, so division is the preferred propagation technique.

CULTURE

Lady palms are rather slow growing, so they need a reasonably well-drained potting mix that won't suffer excessive decomposition. It helps to have some sphagnum peat in the mix, usually with pine bark, sand, and leached perlite. I like to let the mix become somewhat dry, then irrigate thoroughly. The mix should be limed to a pH of no more than 6.0.

Light levels range from 2,500 to 6,000 f.c. (27 to 64.8 klux). In warm climates I prefer the darker end of that range, usually 73% shade. The palms can be grown in full sun, though color will tend to be somewhat pale, and foliage may burn on occasion.

For good, continuous growth it is best to maintain a temperature of 60F (16C) or higher. *R. excelsa* is rather heat tolerant. *R. humilis* prefers cooler conditions, however, so it is more popular in California and the Mediterranean and less common in Florida. Preferred temperatures are between 50 and 80F (10 and 26.7C), though the palms tolerate temperatures as low as 18F (-7.8C) and as high as 90F (32.2C).

NUTRITION

Due to slow growth rates, rhapis palms have modest fertility requirements. Constant liquid feed at 150 ppm nitrogen from 9-3-6 or 24-8-16, with a leach every two to three weeks, is successful. Incorporated or top-dressed slow-release, coated fertilizers, such as 19-6-12 or 18-6-8, at the medium rate also work well. Traditionally, granular fertilizers can also be top-dressed on larger containers, though the slower-release formulations are preferable.

Iron and magnesium should be emphasized in *Rhapis* fertilization programs. These trees have fewer problems with potassium deficiency than many other palms. The fertilizers and irrigation water should be low in fluoride. It helps to have a source of phosphorus available in the mix when lady palms are first potted.

DISEASES

Fortunately, relatively few diseases affect rhapis palms. There are eight to 10 reported fungal leaf spots, though they are fairly uncommon in nursery production. Sprays of Dithane (mancozeb) control most spots that might develop. *Cercospora raphisicola* causes tiny, brown leaf spots about one-sixteenth inch

(2 mm) across. It takes four to eight weeks for the spots to develop after infection. The spots may develop into dark, circular spots with a yellowish halo. Sprays of Cleary's 3336 or Domain (thiophanate methyl) should be helpful against *Cercospora*.

Rhizoctonia occasionally causes crown rot in rhapis palms, especially in warm and wet conditions. Try to dry out the plants if this happens, and use a relatively open, well-aerated and -drained soil mix. Drenching with Chipco 26019 (iprodione) also helps control *Rhizoctonia*.

INSECT AND MITE PESTS

Insect pests are relatively few, though scale insects are occasionally a problem. The scales, generally found on older leaves, may cause leaf drop and reduction in leaf color after a while. Sprays of Dycarb (bendiocarb) or one of the more gentle horticultural oils help control them. Granular Di-syston (disulfoton) is also very good.

The leaves of *Rhapis excelsa* are generally too tough and leathery for spider mites to penetrate, so mites are rarely encountered on *R. excelsa* varieties. The Thailand, or Thai Dwarf, variety is attacked by spider mites from time to time. Several effective miticides are registered, including insecticidal soaps, Talstar (bifenthrin), Pentac (dienoclor), and Avid (abamectin). Otherwise, *Rhapis* is generally pest free.

Fluoride usually comes from either irrigation water, unleached perlite, phosphate fertilizers, or some types of dolomite.

DISORDERS

The most serious disorder of rhapis palms is chlorosis. Significant yellowing of the newer foliage occurs, sometimes with green veins remaining. The older leaves may also become chlorotic, only less so. The ultimate cause is probably lack of iron, which may be brought about by excessive moisture, poor soil aeration, root disease, cold soil, or high pH. Even just drenching or top-dressing with chelated iron sometimes corrects the chlorosis if the cultural condition causing the deficiency is not too severe. If the iron drench doesn't work, then go back and try to figure out and fix the fundamental problem.

I haven't seen it documented in the literature, but rhapis palms are fluoride sensitive. Brown leaf tips frequently develop in interior installations and sometimes in nurseries, as well. Fluoridated city water is especially a problem. Try to use low-fluoride irrigation water and low-fluoride fertilizers. Maintain adequate soil calcium levels and avoid excessive heat and moisture stress. If leaf tips are very badly burned, the problem may be elevated soluble salts rather than fluoride.

Chlorosis tends to develop in lady palms when the pH is above 6.0. Try to keep soil pH in the upper 5 range and maintain good calcium levels. Help keep iron available by allowing the soil to dry out fairly well before irrigating again. Potting soil dries out from the top down, so try to let the soil dry almost to the bottom of the container, but not quite, before watering. Make at least 40% of the potting soil from components that will do not break down physically, such as sphagnum moss, perlite, polystyrene beads, and coarse sand.

TRICKS

The rather tough leaves of rhapis palms are not very conducive to foliar absorption of nutrients. Nutritional adjustments are usually best made via soil applications. In my experience rhapis palms do not like a lot of manganese in the feed program. They need some manganese, like all plants, but higher manganese may be associated with a greater frequency of chlorosis. Try to use either rainwater or good, low-fluoride well water when growing the dwarf rhapis or a rhapis in a large container. It helps to drench newly potted divisions with a high-phosphate starter fertilizer to achieve faster rooting of the transplant.

INTERIOR CARE

Lady palms are generally very durable indoor plants. They seem to prefer cooler interior environments, from 50 to 72F (10 to 22C). They tolerate 45F (7C) without injury. Bright indirect light is best, the ideal light level probably 250 to 300 f.c. (2.7 to 3.2 klux). *Rhapis* tolerates 75 to 100 f.c. (0.8 to 1.1 klux) indoors, but higher light is better. It maintains a good appearance up to 6,000 f.c. (64.6 klux).

Lady palms typically need irrigation two to four times per month, depending on soil mix, age, and conditions. Let them get reasonably dry before irrigating. Fertilize very lightly, preferably with a little iron in the liquid feed once in a while. It is usually best not to fertilize in winter.

Watch for fluoride-induced tip burn. Keep an eye out for scale and mite infestations. Older leaves of *R. humilus* need to be trimmed periodically, while *R. excelsa* requires virtually no pruning.

REFERENCES

Blessington, T.M., and P.C. Collins. 1993. *Foliage Plants: Prolonging Quality,* 178-180. Batavia, Ill.: Ball Publishing.

Chase, A.R. 1987. *Compendium of Foliage Plant Diseases,* 19. St. Paul, Minn.: The American Phytopathological Society.

Dransfield, J. 1985. Foliage Plants Update. *Greenhouse Manager* (January): 16.

Kraft, K. March 1996. Telephone conversation with author.

McKamey, L. 1984. Oriental Elegance. *Interior Landscape Industry* (August): 51-56.

Stewart, L. 1994. *A Guide to Palms and Cycads of the World,* 171-172. Sydney, Australia: Angus & Robertson.

Watkins, J.V., and T.J. Sheehan. 1975. *Florida Landscape Plants,* 46. Gainesville: The University Presses of Florida.

RUBBER PLANT

(*See* Ficus elastica)

SCHEFFLERA ACTINOPHYLLA

Umbrella Tree

HABITAT

Schefflera actinophylla, the Queensland umbrella tree, has recently undergone a name change. The Latin name used to be *Brassaia actinophylla*, the genus named for English botanist W.P. Brass and the species for leaflets in a radial arrangement. Taxonomists have now reversed themselves, so the Latin name is back to *Schefflera actinophylla*, after eighteenth-century botanist A. Scheffler. *Schefflera arboricola* varieties will be handled separately in the next variety section. I will use the trade name "schefflera" here to refer specifically to *S. actinophylla*.

Fig. 37 *Schefflera actinophylla* Amate, grown by the author. (Photo by the author)

A native of Queensland, Australia, schefflera grows as a medium-sized tree, typically in USDA Zones 9 and 10. In its native habitat it is frequently found growing in porous mulch or humus beds, as well as epiphytically in crotches of trees and rock formations. The plant retains some of those epiphytic characteristics when grown in containers, as well.

USES

Scheffleras are grown as multiple-plant specimens in containers ranging from 6 to 17 inches (15 to 42 cm). With multiple plants in the pot, it looks like a full, large-leaved shrub, though it is really a tree and will ultimately grow with a more upright, tree habit. Scheffleras are durable interior plants if kept in good potting soil and irrigated reasonably well. Scheffleras are popular landscape plants in Florida, Hawaii, and California, typically being medium-sized trees of about 20 feet (6 m). Dark red flowers are produced in summer, with black seed heads quickly following in late summer or fall.

VARIETIES

Most scheffleras are grown from seed collected in the landscape, so the garden-variety schefflera is not a named cultivar. The registered trademark variety Amate is named after a California nurseryman, Archie Amate. This selection, typically produced from tissue culture, was originally prized for its resistance to *Alternaria* and other leaf spot diseases. Experience has also shown it to be a sturdy, thicker variety of schefflera, one that has resistance to spider mites and sells for a good price in the marketplace. Variegated examples of *S. actinophylla* are occasionally encountered, though the variegation tends to be very unstable.

PROPAGATION

Scheffleras can be produced from cuttings and air layers, but the garden variety is almost always grown from seed. Seeds are collected in late summer and can be stored refrigerated after cleaning. Seeds should be just barely covered in a peat-based medium, either in beds, flats, or cell trays. Even planting depth will result in more even germination and plant size. Germination percentage is usually 80% or better, and the seedlings typically come up in three weeks or so. For fullness try to have at least 12 plants per pot ultimately, though some growers use many more than that.

Amate is usually produced from tissue-culture liners potted directly into the finishing container.

CULTURE

Scheffleras are generally grown on ground cover in shadehouses, though they are seen on benches in greenhouses in more temperate climates. Fifty-five percent shade, equivalent to 6,000 f.c. (64.5 klux), is ideal. When grown under less light, the plants get deep, shiny, green foliage, though new growth tends to be weak and stretched, with extended internodes. They can be started in full sun and shaded when they are one-half to two-thirds marketable size.

Scheffleras do best in a high-porosity growing medium with not more than 40% peat. Some sand in the mix is helpful when growing in shade-houses, as plants tend to blow over and become deformed and damaged as they approach maturity. It is usually necessary to drench with fungicides at potting, and plants showing damping-off may require a second drench (see "Diseases").

Scheffleras tolerate a wide range of temperatures, from 35 to 105F (2 to 40C), with the optimum range being 60 to 90F (16 to 32C). Amate is usually produced from tissue-culture liners, using two plants per pot for an 8-inch (20-cm) pot, three plants for a 10-inch (25-cm), and four plants for a 14-inch (35-cm). Plants in 10-inch containers are generally marketable when they reach 36 to 42 inches (0.9 to 1.05 cm). When a plant gets large in the container, it can be difficult to wet with overhead irrigation—it's not called the umbrella plant for nothing. Growing time for a 10-inch plant from a liner is about six months.

TABLE 23 **Leaf analysis rating standards for *Schefflera actinophylla***

Nutrient (%)	Very low	Low	Medium	High	Very high
Nitrogen	<2.0	2.0–2.4	2.5–3.5	3.6–4.5	>4.5
Sulfur	<0.15	0.15–0.20	0.21–0.80	0.81–1.20	>1.20
Phosphorus	<0.15	0.15–0.19	0.20–0.50	0.51–0.80	>0.80
Potassium	<1.80	1.80–2.20	2.25–4.00	4.05–5.00	>5.05
Magnesium	<0.20	0.20–0.24	0.25–0.75	0.76–1.00	>1.00
Calcium	<0.80	0.80–0.99	1.00–1.50	1.51–2.50	>2.50
Sodium			<0.21	0.21–0.50	>0.50
(ppm)					
Iron	<40	40–49	50–300	301–500	>500
Aluminum			<101	101–250	>250
Manganese	<40	40–49	50–300	301–500	>500
Boron	<15	15–19	20–60	61–100	>100
Copper	<6	6–9	10–60	61–200	>200
Zinc	<15	15–19	20–200	201–400	>400

Sources: Institute of Food and Agricultural Sciences, Apopka, Florida; Dr. Benjamin Wolf, Fort Lauderdale, Florida.

Notes: Common names include schefflera, umbrella tree, and arboricola. Sample of most recent fully mature leaves, no petioles.

NUTRITION

Scheffleras do best with relatively high amounts of fertilizer in order to achieve the best color, growth rate, and disease resistance. N-P$_2$O$_5$-K$_2$O ratios of 3-1-2 or similar are generally used at the high rate. The trees also do well with a 4-1-2 or 5-1-2 ratio. They do not have a high requirement for trace elements, and I generally do not recommend incorporating minor elements into the potting medium for scheffleras. They occasionally develop magnesium deficiency symptoms in the nursery, as indicated by broad, marginal yellowing in the older leaves. Foliar applications of trace elements are generally not needed and not recommended on scheffleras.

DISEASES

The best known and most serious disease of scheffleras is *Alternaria panax*. The fungus causes a series of large, brown, irregularly shaped leaf spots, which can spread quickly. Growers refer to the complex as *Alternaria,* though there are actually other fungi that cause very similar symptoms. The disease can be prevented by keeping the foliage dry or at least minimizing the amount of time the foliage stays wet. Many schefflera growers use preventative sprays of Dithane (mancozeb) or Chipco 26019 (iprodione). Iprodione can be occasionally phytotoxic, especially on Amate; however, when used with care, it is quite effective against the disease.

Phytophthora parasitica causes leaf spots very similar in appearance to those from *Alternaria,* as does *Colletotrichum gloeosporioides.* Dithane (mancozeb) or Aliette (fosetyl-aluminum) are helpful against *Phytophthora. Colletotrichum* can be controlled with sprays of Zyban (thiophanate methyl and mancozeb) or either active ingredient individually.

Damping-off is rather common, especially in plants less than 3 inches (7.5 cm) tall. Liners exposed to heavy or frequent rains within a month after potting may suffer serious losses.

Damping-off is frequently caused by *Pythium* or *Rhizoctonia,* though other pathogens may be involved as well. Soil drenches of liners and young plants with Banrot (etridiazole plus thiophanate methyl) are effective, as is the combination of Chipco 26019 (iprodione) plus Subdue (metalaxyl). Avoiding excessive soil moisture is also a big help.

The bacterium *Xanthomonas,* which occasionally causes a small, corky leaf spot, can be controlled with Aliette (fosetyl-aluminum) sprays as well as high fertility. Brown blisters resulting from edema may appear on the older leaves of mature plants at times. Edema is a physiological disorder caused by moisture fluctuations, specifically too much moisture within the plant. Finally, the fungus *Sphaerobolus*, the popcorn or bird's-nest fungus, may occasionally appear as small, brown disks on the underside of the leaf. They may easily be mistaken for scale insects.

 Alternaria sporulates better under high light.

INSECT AND MITE PESTS

The two-spotted mite *(Tetranychus urticae)* is by far the most common schefflera pest. The mite proliferates in hot, dry weather, though mite injury, characterized by leaf distortion and small, white specks, can show up anytime. Most schefflera growers spray preventatively for mites, using Kelthane (dicofol), Pentac (dienochlor), Ornamite (propargite), or Avid (abamectin). For good mite control you must get good coverage, which involves having enough spray pressure to reveal the undersides of the leaves. Wetting agents or surfactants are also helpful.

Aphids occasionally show up on new foliage, especially in spring. Orthene (acephate) sprays are generally used to control them. Scale and mealybug infestations are relatively rare, but they do occur. Sprays of Dycarb (bendiocarb) are effective. I should note here that most of the pesticides on the market have been reported as being phytotoxic on scheffleras at one time or another. The plants are rather susceptible to spray injury, so apply chemicals with care and not in hot weather.

DISORDERS

Scheffleras can turn pale when soil temperature drops below 60F (15.5C). Color usually returns within two weeks once the soil warms up again. Phytotoxicity can result from almost any chemical if too much spray pressure is applied too close to the plant. High pressure injures the soft, delicate, immature foliage, resulting in distortion of the compound leaves as they grow out. Try to spray with enough pressure to turn the leaves over for mite control, but be far enough away from the plants to not injure the immature foliage with spray pressure.

High soluble salts can cause root death, though scheffleras are fairly tolerant of high salts. Iron and manganese toxicity are not uncommon in scheffleras. The plants become stunted, and the older leaves show pronounced dark purple to black veins, with chlorosis in between. Trace element toxicity can occur when minor elements are incorporated into the potting mix for young plants, when pH drops too low, or when root rot occurs in young plants. Liming may help alleviate the problem, as can fungicide drenches if roots are diseased. Boron toxicity is indicated by narrow chlorotic margins of older leaves, and the edges become burned in appearance.

TRICKS

B-nine (daminozide) is very effective for controlling the height of scheffleras. It can reduce stretch of the petioles when growing in low light. B-nine is also used at times to keep plants of finished size from becoming overgrown before they are sold. It also improves the color.

The combination of Chipco 26019 and Subdue (iprodione and metalaxyl) is usually more effective as a drench than other fungicide combinations on scheffleras. Neither Dycarb (bendiocarb) nor Orthene (acephate) should be drenched on them. Avoid getting the spray gun too close to the plants when spraying. Wettable powder formulations are generally less phytotoxic on this plant than emulsifiable concentrates or oil-based sprays. Sprays of common household white vinegar diluted 1:100 are effective against *Xanthomonas* leaf spot.

When scheffleras are happy, the leaves develop a very high shine, or gloss. Use this fact when evaluating how well you are growing the plants. They tend to droop slightly on warm afternoons, even with good soil moisture, so don't worry about it; resist the temptation to water.

INTERIOR CARE

Scheffleras make reasonably good interior plants when maintained in high-porosity potting media and in the absence of overwatering. It is better to let them wilt slightly before watering than to risk irrigating excessively, which rots roots and kills plants. However, when plants are stressed from lack of moisture, spider mite activity can be heightened.

Scheffleras can be kept in dark conditions up to 30 days without severe negative effects. If not overwatered, scheffleras can last indefinitely under interior lighting of 150 f.c. (1.5 klux) or greater. Brighter light is better, but don't keep the plants under lights more than 16 hours per day.

Inspect the plants periodically for mites. Ethylene sensitivity is moderate. Fertilize three to four times per year with 1 teaspoon of 20-20-20 per gallon (3.8 l).

REFERENCES

Ben-Jaacov, J., R.T. Poole, and C.A. Conover. 1982. Effects of Long-Term Dark Storage on Quality of Schefflera. *Hortscience* 17 (3): 347-349.

Braswell, J.H., T.M. Blessington, and J.A. Price. 1982. Influence of Production and Postharvest Light Levels on the Interior Performance of Two Species of Scheffleras. *Hortscience* 17 (1): 48-50.

Chase, A.R. 1995. What Ails It? *Landscape and Nursery Digest.* (February): 34-35.

Chase, A.R., and R.T. Poole. 1987. High Fertilizer Rates Reduce Severity of Xanthomonas Leaf Spot on Scheffleras. *Folage Digest.* (December): 4-5.

Colijn, A.C., and R.K. Lindquist. 1985. Brassaia and Schefflera. Arc-A Foliage Plant Research Note RH 1983-C.

Osborne, L.S. 1983. Don't Drench Plants with Pesticides Meant to be Used as Foliar Sprays. *Nursery Notes.* 1-3.

Poole, R.T., and C.A. Conover. 1984. Tolerance of Schefflera Growing Indoors to Soluble Salts. Arc-A Research Report RH-84-4.

SCHEFFLERA ARBORICOLA

Dwarf Schefflera

HABITAT

The more than 150 species of *Schefflera*, members of the family Araliaceae, are generally native to the tropics of the Far East and Australia. *Schefflera arbori-cola*, commonly known as the dwarf or Hawaiian schefflera, is native to Taiwan, where cuttings are frequently fed to goats. The dwarf schefflera has a palmate pattern of approximately eight radiating leaflets, which may appear dull or shiny. The seeds were first introduced to the United States foliage trade by Ron Huroff of California in the late 1960s.

S. arboricola seems to come from a climate somewhat more variable than that of *S. actinophylla*. Like *S. actinophylla*, *S. arboricola* can grow somewhat epiphytically, with numerous aerial roots. It is generally found at lower altitudes and also near the sea. The plant is much more salt tolerant than people realize.

Fig. 38 *Schefflera arboricola* from a stock farm in Central America. (Courtesy of Bill Lewis)

USES

S. arboricola has been in the U.S. foliage trade since the early 1970s. Liners with many seeds per pot are used to grow it in bush form. Pot sizes range from 3 to 10 inches (7.5 to 25 cm), with all sizes in between. Standard tree forms are produced from air layers in 8- to 10-inch (20- to 25-cm) and occasionally larger containers. Braided standards with three to four stems per pot are also grown, in a fashion similar to *Ficus benjamina*. This can be especially attractive with variegated types whose stems also show variegation. Bonsai forms are also produced.

VARIETIES

There are at least three green types of dwarf schefflera. Jakarta Jewel has more pointed foliage and a slightly duller green color. The U.S. market generally prefers a more rounded and deeper green leaf, and though it doesn't have a name, the rounder-leaved version is usually considered the standard dwarf schefflera in the trade. The cultivar Renate has a leaf lobed near the tip, somewhat resembling a footprint.

Numerous variegated cultivars exist, including Gold Capella, which has deep green, shiny, oval leaves with intensely contrasting yellow variegation. Goldfinger has more narrow variegated leaves than Gold Capella. Jacqueline is somewhat similar to Gold Capella, though the variegation is more gold in color and tends to concentrate along the leaf margins. Some other *S. arboricola* varieties in the trade include Hong Kong, Trinette, Covette, Worthy, and Henrietta. The seedlings from Henrietta make a nice dwarf nonvariegated plant.

PROPAGATION

S. arboricola may be propagated from seed, tip cuttings, stem cuttings, leaf-and-eye cuttings, and air layers. Seedling liners are frequently grown in cell trays, with at least 20 seedlings per pot, sometimes as many as 40 or 50. The tan-colored seeds are harvested and cleaned from the golden fruits in the fall. I do not recommend incorporation of trace elements into the medium in liner seed production.

The various types of cuttings root easily under mist or fog, generally in two to three weeks. Cuttings taken from the more basal portion of the stem tend to develop more roots and longer shoots and to branch better than tip cuttings. One research paper suggests misting for 30 seconds every six minutes, but that sounds like too much to me. I used to mist five tip cuttings approximately 8 inches (20 cm) long in a 6-inch (15-cm) pot and sell them right off the mist bench. Five seconds of mist every six to 10 minutes should be enough, depending on greenhouse conditions. Leaf-and-eye cuttings work well if stock is limited, though cultural practices during rooting must be watched carefully, as disease pressure is greater.

Air layers from field-grown stock plants may be easily propagated using moist sphagnum moss, aluminum foil, and twist ties. This type of propagation is generally reserved for standards and braids.

High bicarbonates in irrigation water tend to remove calcium from soil.

CULTURE

S. arboricola is rather tolerant of a wide variety of growing conditions. The plant tolerates between 35 and 105F (1.67 and 40.5C). With 77F (25C) the optimum temperature, the best growing temperatures are generally maintained between 65 and 90F (18.3 and 32.2C). The growth rate may be slower with temperatures above 90F (32.2C). Quality begins to drop as temperatures dip below 50F (10C). It is best to keep the soil temperature above 65F (18.3C), as plants begin to discolor and droop as it falls below 60F (15.6C).

Dwarf scheffleras are fairly tolerant of most types of potting media, as long as they are irrigated reasonably competently. Sedge peat may be mixed with bark or wood chips and about 10% sand by volume. Rock wool can also be used as long as the percentage is not too high.

Opinions differ on the best light levels for producing *S. arboricola*. One recommendation is 3,000 to 5,000 f.c. (32.4 to 54 klux), though I prefer 47 to 55% shade (approximately 6,000 f.c. or 64.6 klux). I haven't steered you wrong yet, have I? Stems are sturdier and foliage more shiny under higher light. Many growers produce *S. arboricola* in direct sun and finish it for a few weeks under 63% shade. Pruning is generally not necessary in small pots, but long shoots may occasionally be trimmed to keep larger plants bushy.

NUTRITION

The fertility requirements of dwarf schefflera are medium to high in comparison to other foliage plants. The typical 3-1-2 ratio is commonly used, from either granular topdress fertilizers, slow-release, coated fertilizers, or constant liquid feed at 200 ppm nitrogen. Granular 12-6-8 with magnesium and trace elements has been a popular *S. arboricola* fertilizer for many years, using about 2 tablespoons per 10-inch (25-cm) pot as a topdress.

S. arboricola is rather sensitive to high ammonia levels. It helps to have molybdenum in the program to help convert excess ammonia to nitrate. *S. arboricola* is probably more sensitive to ammonia than to soluble salts in general. Nitrate forms of nitrogen grow a good, sturdy *S. arboricola* in the winter in northern climates.

S. arboricola is tolerant of a fairly wide range of fertility levels, though at higher fertility it is somewhat more disease resistant. The plant likes sulfur, and its color will be much better if you have good levels of sulfur in the fertilizer program. I do not recommend incorporation of trace elements into the medium for *S. arboricola,* as trace element toxicities of iron and manganese can occur. Limit foliar sprays of iron compounds.

DISEASES

The most common foliar disease of *S. arboricola* is *Pseudomonas cichorii*. Foliage that is frequently wet will develop small, water-soaked, marginal leaf spots that become large and turn purplish to black. Infected leaves usually drop.

High feed rates and high nitrate nitrogen discourage *Pseudomonas* somewhat. Keep the foliage dry to the extent possible, and spray with Dithane (mancozeb) and a copper fungicide.

A less common bacterial leaf spot is caused by *Xanthomonas campestris* pv. *hederae,* the same bacterium that attacks English ivy. Tiny, yellow to tan leaf spots form, and they may coalesce into slightly larger tan spots. Sprays of Aliette (fosetyl-aluminum) at 2 pounds per 100 gallons (0.9 kg per 380 l) are helpful, as are mancozeb and copper. Don't apply copper and Aliette close together in a spray rotation, however, or phytotoxicity may result. Increasing fertility also reduces incidence of *Xanthomonas.*

Alternaria panax, another common leaf spot, is indicated by small spots that begin yellow or tan and ultimately turn medium brown. Again, high feed rates are helpful, as is keeping foliage dry. This disease kills stems if plants are cut back when inoculum level is high. Sprays of Chipco 26019 (iprodione) or Dithane (mancozeb) are effective.

As with regular *S. actinophylla, S. arboricola* is attacked by *Pythium* root rot if the medium stays too wet. If you pull gently on a small root, the cortex of the root will slough off, leaving only the string of the internal root tissue behind. Dry the plants out, then drench with either Subdue (metalaxyl), Aliette (fosetyl-aluminum), or Truban (etridiazole).

Rhizoctonia can be very troublesome in seedling beds or as a stem rot of cuttings in propagation. Seedlings turn a tan color at the base of the stem, and plants fall over by the hundreds. Stems of cuttings in propagation turn medium brown, and the cuttings ultimately die. A drench with Chipco 26010 (iprodione) usually stops *Rhizoctonia* when used in conjunction with good moisture management. Incidentally, nematodes rarely affect *S. arboricola.*

INSECT AND MITE PESTS

Like most plants in the Aralia family, *S. arboricola* is frequently visited by the two-spotted mite, *Tetranychus urticae.* Leaves become speckled, and the plant may develop a grayish appearance, with more leaf loss than normal. Sprays of Pentac (dienoclor), Avid (abamectin), or Kelthane (dicofol) are effective. Predator mites and other IPM (integrated pest management) methods are also effective. A barely visible, tiny mite also attacks the brand-new, emerging foliage of *S. arboricola,* causing purplish or reddish leaf spots. Sprays of Thiodan (endosulfan) usually control the mites.

Aphids may be found on new foliage, especially in spring or whenever else growth is abundant. Orthene (acephate) is commonly used, but there are many other effective controls. Mealybugs are not common, though their white, cottony masses are occasionally observed on the undersides of older leaves and on the petioles. Granular Di-syston (disulfoton) controls them, as do sprays of Dycarb (bendiocarb) or horticultural oils. Scale insects, including Florida wax scale, cottony cushion scale, and nigra scale, are an occasional problem on dwarf schefflera. Controls are similar to those for mealybugs. Cygon (dimethoate) may be used, but with caution. Thrips occasionally cause scarring and distortion of new foliage. Sprays of Mavrik (tau-Fluvalinate) or Thiodan (endosulfan) usually control them. Leafminer, a relatively uncommon pest of foliage plants, can also be troublesome, especially when growing outdoors or in the landscape. Sprays of Avid (abamectin) usually stop leafminers.

The bird's-nest fungus, *Sphaerobolus*, causes small, brown disks on the undersides of older leaves. The disks look much like scale insects, but aren't. The problem is related to the wood products in potting soil or mulch, and insecticides do nothing to it.

DISORDERS

As I indicated earlier, toxicities of iron and manganese are fairly frequent in *S. arboricola*. I have seen this only on plants produced from seed, never from cuttings or air layers. The symptoms include an odd yellow mottling in the older foliage, frequently with purplish inclusions within the leaf. The symptom is often associated with root rot, and I suspect overwatering contributes to the problem. Affected plants are stunted and leaves drop. Liming may help, but some of the plants with this condition may never recover. Also, high soluble salts tend to cause brown leaf edges, especially in older foliage. The fungicide Spotless (triazol) may cause phytotoxicity at high rates on this plant.

S. arboricola droops when soil oxygen is low or ethylene is present. If you get heavy rains just after potting seedling liners, huge losses from disease can occur. If soil temperature drops below 60F (15.6C), plants become pale, yellowish, and somewhat sickly looking. When dwarf scheffleras are too dry, the foliage turns somewhat grayish, and older leaves turn yellow and drop.

TRICKS

S. arboricola is much more salt tolerant than the industry realizes. I have seen it growing right on the beach on Grand Cayman and have recommended the plant frequently for coastal landscapes in the Caribbean and Hawaii. It is not particularly tolerant of high winds, but it handles the salt fabulously.

Sprays of vinegar at 1 gallon per 100 gallons (3.8 1 per 380 1) of water are effective in controlling foliar *Xanthomonas*. When growing in full sun, plant color can be made to approach that of shade-grown plants by spraying with a mixture of magnesium sulfate and manganese sulfate. Sprays of the growth regulator B-nine (daminozide) are very effective in improving plant color in *S. arboricola* and in avoiding excessive growth. B-nine sprays help keep plants from becoming overgrown before they can be sold. When spraying dwarf scheffleras, concentrate the spray on the underside of the foliage.

Dwarf scheffleras don't like cold irrigation water, which reduces plant quality even if culturally everything else is perfect. Arboricolas do not do well with low soil pH; liming can do wonders for older plants with low soil pH.

INTERIOR CARE

S. arboricola is more shade tolerant than regular *S. actinophylla,* and it tends to stretch much less in the interior environment. It is quite adaptable to interior light levels as long as it is not overwatered. The best day temperatures are 65 to 75F (18 to 24C), and the plant is certainly best kept above 60F (16C). An occasional draft of 45F (7C) can be tolerated without incident.

Minimum interior light levels of 75 to 250 f.c. (0.8 to 2.7 klux) are recommended, but more light is better. Apply only minimum fertilization, unless light levels are decent. When maintained indoors under good light, occasional fertilization using 2 teaspoons 20-20-20 and 1 teaspoon magnesium sulfate per gallon (3.8 1) will maintain good color and growth.

Leggy growth may be trimmed to keep *S. arboricola* plants compact and symmetrical. Avoid letting the plants sit in a wet saucer, or quality will quickly diminish.

REFERENCES

Blessington, T.M., and P.C. Collins. 1993. *Foliage Plants: Prolonging Quality,* 188-190. Batavia, Ill.: Ball Publishing.

Braswell, J.H., T.M. Blessington, and J.A. Price. 1982. Influence of Production and Postharvest Light Levels on the Interior Performance of Two Species of Scheffleras. *Hortscience* 17 (1): 48-50.

Chase, A.R. 1989. Fungicides for Control of Leaf Spots of Foliage Plants. *Florida Nurseryman* (February): 67-68.

Chase, A.R. 1990. Nitrogen, Phosphorus and Potassium Rates Affect Xanthomonas Leaf Spot of Schefflera. *Foliage Digest* (September): 7-8.

Chase, A.R. 1995. What Ails It? Common Diseases of Schefflera and Dwarf Schefflera. *Landscape and Nursery Digest* (February): 32-34.

Chase, A.R., and R.T. Poole. 1987. High Fertilizer Rates Reduce Severity of Xanthomonas Leaf Spot of Scheffleras. *Foliage Digest* (December): 4-5.

Conover, C.A., A.R. Chase, and L.S. Osborne. 1983. Brassaia and Schefflera. *Nurserymen's Digest* (September): 90-93.

Deneve, Bob. April 1996. Telephone conversation with the author.

Ellison, D.P. 1995. *Cultivated Plants of the World*, 490-492.

Lorenzo-Minguez, P., R. Gabriels, I. Impens, and O. Verdonck. 1985. Response of Gas Exchange Behavior on *Schefflera arboricola* to Air Humidity and Temperature. *Hortscience* 20 (6): 1060-1062.

Poole, R.T., and C.A. Conover. 1989. Growth of *Dieffenbachia* 'Camille' and *Schefflera arboricola* With Gro-Prod as a Potting Ingredient. *Foliage Digest* (November): 7-8.

SPATHIPHYLLUM

Peace Lily

HABITAT

Spathiphyllum, commonly known as the peace lily, is one of the more exceptional interior plants because it does very well under low-light interior conditions and also flowers in the interior. The genus name means "leaf spathe." Usually white-flowered, the peace lily has a hood-shaped spathe sheltering the spadix, the erect part of the flower containing the pollen and the seeds.

There are about 35 known species of *Spathiphyllum*, about 30 of which are native to Central and South America, while two are found in the Malay archipelago, completely on the other side of the world. One species is found in two places separated by the vast Pacific, an island off Costa Rica and in the Philippines, which gives excellent credence to the theory of continental drift.

256

The plant generally grows naturally in warm, humid tropical rain forests, where it is densely shaded by surrounding jungle vegetation. The peace lily therefore does well in moist, low-light situations.

Fig. 39 *Spathiphyllum* Mauna Loa Supreme in full flower. (Photo by the author)

USES

Some spath varieties are only about 4 inches (10 cm) tall, whereas others reach 10 feet (3 m). Foliage growers generally produce them in pot sizes ranging from 4 to 21 inches (10 to 52.5 cm), with 6- and 10-inch (15- and 25-cm) pots making up the bulk of U.S. production. Several plants per pot are grown in a clump for fullness. Peace lilies are grown in full-sized pots, as well as azalea pots and bulb pans for the smaller varieties.

For interiorscapes spathiphyllums provide a tropical look, with the added benefit of the exotic-looking white flowers. Blooming generally starts between February and April, with the peak bloom season running from about April to September. Some plants flower intermittently during the winter. Mass plantings in beds are popular, as are individual potted specimens. Smaller spathiphyllums are frequently planted indoors under large trees.

VARIETIES

Numerous cultivars exist in the trade today, while many others have faded from the scene in the last 10 years. The cultivars may be divided into three classes: large, medium, and small varieties. Mauna Loa is one of the older, more widely grown, large varieties. It grows rapidly and likes frequent irrigation. Mauna Loa is a hybrid of a hybrid; therefore, it has irregular bloom characteristics, producing both large and small flowers, which may be green, white, or a mixture of both. Supreme, an improved version of Mauna Loa, has a higher percentage of all-white flowers. Sensation is the largest common cultivar, reaching about 5 feet (1.5 m) tall. It resulted from a cross of Mauna Loa Supreme and Fantastica.

Lynise is an attractive type resistant to wilt when dry. My friend Bond Caldwell created it by crossing Floribundum with Tasson. Deneve is an attractive variety with less shiny foliage that appears more serrated and pleated.

The medium-sized varieties are generally produced in containers between 6 and 10 inches (15 and 25 cm). Among them, Tasson has been popular for at least 15 years. Coming from a cross of Mauna Loa and Wallisii, it has a compact habit, somewhat rounded leaves, and numerous medium-sized flowers. Viscount, similar to Tasson and with similar lineage, is slightly larger and taller and tends to be more uniform genetically.

Of the smaller varieties, Petite has been around for a number of years. It is especially good for bulb pans and azalea pots. Petite suckers well and is a free bloomer. Wallisii has more narrow leaves than Petite and a somewhat lower growing habit. Starlight is similar to Wallisii, though slightly larger and more vigorous. Numerous other spath varieties are grown in the trade, including a number of new European cultivars.

PROPAGATION

Mauna Loa is usually grown from seed, though the majority of the other cultivars today are produced by tissue culture. To generate seed and to cross spathiphyllum flowers, pollen is collected and used either right away or stored up to a week in the refrigerator. Plastic film canisters are handy for collecting pollen. The pollen is applied to the pearly white flower knobs with a small paintbrush. Once the flower knobs become dark, they are no longer receptive to pollen. Seeds will swell after about a month, becoming ripe when they turn tan about three to four months later.

Tissue-culture liners are normally produced by various laboratories as multiple-plant clumps in cell packs. Quality and uniformity are generally very good, though there is usually a small percentage of liners genetically distinct from the intended variety. Spathiphyllums were among the first plants to be produced from tissue culture in the United States.

CULTURE

Peace lilies are generally grown in fairly heavy shade, frequently 70 to 73% shade in winter, 80% shade in summer. Typical light levels are therefore 1,500 to 2,500 f.c. (16.2 to 27 klux). Plants grown under excessive light are pale and unattractive. Spathiphyllums are more cold tolerant than they appear. The plants tolerate from 40 to 100F (4.4 to 37.8C), though the preferred temperature range is 65 to 90F (18.3 to 32.2C). Generally they don't like wind, and cold wind causes older leaves to become necrotic.

Peat-based potting media are used, of either sedge peat, sphagnum peat, or a combination. Bark, wood chips, sawdust, and styrofoam beads are also commonly used in various percentages for spathiphyllum production. The mix should have good aeration but a fairly high moisture-holding capacity at the same time, as spathiphyllums are rather thirsty plants. The mix should be limed to a pH between 6 and 7 for most situations. Use plenty of dolomite because the plants have a high magnesium requirement. They are also rather sensitive to soluble salts, so avoid incorporation of high rates of fertilizer, unless it is a very gentle formulation.

NUTRITION

Peace lilies are fairly heavy feeders and are usually grown with a 3-1-2 ratio of N-P_2O_5-K_2O. Smaller pot sizes are generally produced with liquid fertilizer. The

TABLE 24 Leaf analysis rating standards for *Spathiphyllum*

Nutrient (%)	Very low	Low	Medium	High	Very high
Nitrogen	<3.00	3.00–3.29	3.30–4.50	4.51–5.50	>5.50
Sulfur	<0.16	0.16–0.19	0.20–0.50	0.51–1.00	>1.00
Phosphorus	<0.16	0.16–0.19	0.20–0.50	0.51–1.00	>1.00
Potassium	<2.00	2.00–2.29	2.30–4.00	4.01–6.00	>6.00
Magnesium	<0.22	0.22–0.29	0.30–0.50	0.51–1.00	>1.00
Calcium	<0.75	0.75–0.99	1.00–2.00	2.01–3.50	>3.50
Sodium			<0.21	0.21–0.50	>0.50
(ppm)					
Iron	<25	25–49	50–300	301–1000	>1000
Aluminum			<251	251–2000	>2000
Manganese	<25	25–49	50–300	301–1000	>1000
Boron	<20	20–24	25–70	71–100	>100
Copper	<5	5–7	8–100	101–500	>500
Zinc	<18	18–24	25–200	201–1000	>1000

Sources: Institute of Food and Agricultural Sciences, Apopka, Florida; Dr. Benjamin Wolf, Fort Lauderdale, Florida..

Notes: Common names include peace lily, spathiphyllum (mature). Sample of most recent fully mature leaves, no petioles.

larger containers are usually grown with dry fertilizer, either granular or coated, slow-release products. Both 19-6-12 and 18-6-8 are popular blends for spaths.

These plants have higher-than-normal requirements for magnesium, potassium, and boron. Lack of magnesium causes severe marginal yellowing of the older foliage, especially in older plants. Magnesium from sulfate, nitrate, or chelate may be sprayed, drenched, or top-dressed to combat magnesium deficiency. Lack of iron tends to cause a slight veinal chlorosis of the new leaves. Lack of manganese is similar, though the leaves are frequently misshapen. Potassium deficiency appears as necrotic and chlorotic flecks in the older foliage. Low boron or calcium causes longitudinal ribbing in the new leaves and frequently poor flower quality.

DISEASES

By far the most serious disease of spathiphyllums is *Cylindrocladium spathiphyllii,* which was first discovered in Broward County, Florida, in December 1978 from samples collected by my predecessor at A & L Southern Agricultural Laboratories, Wayne Poole. Much has been written about this disease, but here is the short version of what you need to know as a grower. Almost all spath varieties are susceptible, though you will hear otherwise. Variations in disease severity are a function of inocula, and not resistance to this disease. Symptoms include rapid root death, wilt, yellowing of older foliage, and dark brown lesions near the base of the petioles. *Cylindrocladium* kills the plants quickly and is spread by splashing water, flowing water, and handling of plants. The disease is exacerbated by low pH, high temperature, and deep planting.

Dump badly infected plants, then disinfect the bench or ground cover in their vicinity. Treat the remainder with either Terraguard (triflumizole) or Phyton 27 (picro cupric ammonium formate). Grow off the ground to reduce disease spread. The only resistant species is *Spathiphyllum floribundum,* but its resistance is limited.

Pythium and *Phytophthora* also cause root rot in *Spathiphyllum,* though the symptoms are different. *Pythium* is a less severe, less aggressive root rot. *Phytophthora* looks somewhat like *Cylindrocladium* in this plant, though the infection tends to continue to run up the petiole, turning the petioles as well as the roots black. A drench with Subdue (metalaxyl) or Aliette (fosetyl-aluminum) should help against both of these organisms. Foliar *Phytophthora* is also common, especially in rainy weather. Large, irregularly shaped, black lesions form on the foliage. Sprays of Aliette or Daconil (chlorothalonil) or a drench with Subdue should control it, along with a regime of keeping plants dry.

Myrothecium, the other major leaf spot of spath, causes brown, circular lesions that may turn somewhat blackish. Black and white fruiting bodies are frequently observed on the underside of the spot, which may have a white fringe around it. The key is that it is brown and circular, rather than the black, irregular spot of *Phytophthora.* Keep foliage dry and spray with either Daconil (chlorothalonil) or Dithane (mancozeb). *Myrothecium* looks somewhat different on liners and causes the older leaves to blacken. Controls are the same.

Southern blight is occasionally seen on peace lilies in hot weather, with the usual symptoms of white threads of mycelia and little fruiting bodies (sclerotia) that look like mustard seeds. A drench with Terraclor (PCNB) or the insecticide Dursban (chlorpyrifos) helps. Dasheen mosaic virus is not a major problem on *Spathiphyllum,* but it is occasionally seen. The foliage develops yellowish to light green, ghostlike spots. There is no control, but the virus is vectored by aphids. The fact that aphids rarely attacks spath probably explains the rarity of the virus.

I have seen $100,000 worth of spath go to the dump in one week because of Cylindrocladium.

INSECT AND MITE PESTS

Insect problems are relatively few in *Spathiphyllum* production. Caterpillars occasionally chew holes in the leaves, but sprays of either Dipel *(Bacillus thuringiensis)* or Sevin (carbaryl) control them. Snails also occasionally chew holes in foliage, especially in springtime. Their chewing injury tends to be somewhat smaller, but they leave dark, somewhat elongated droppings, as opposed to the round droppings of caterpillars. The molluscicide Grand Slam (methiocarb) is helpful, as are the metaldehyde baits.

Spider mites and scale insects are rarely seen on *Spathiphyllum.* Thrips are an occasional problem, causing some distortion and injury to emerging foliage. The damage is frequently more on one side of the midrib than the other. Sprays of Mavrik (tau-Fluvalinate) usually control them.

DISORDERS

The common fungicide Chipco 26019 (iprodione) is phytotoxic to some spathiphyllum cultivars, both as a spray and a drench, in my experience.

Symptoms include numerous tiny, rust-colored spots on the foliage and a somewhat brittle, cupped appearance to the leaves. Iprodione is a great fungicide, but not on this plant. If soils are especially saturated or cold, plants wilt and lack vigor. High light levels cause pale color, chlorosis, tip burn, and leaf curl.

There is generally no need to spray spathiphyllums more than very occasionally with iron. Iron toxicity, occasionally observed in peace lilies, appears as tiny, gray or black blotches in most of the leaves. The insecticide Orthene (acephate) can be safely used on spathiphyllums, but burn can occur at high temperatures; try to spray when it's below 80F (26.7C).

TRICKS

Color in spathiphyllums growing under high light conditions can be improved by spraying about every two weeks with magnesium nitrate at 1 quart per 100 gallons (0.95 l per 380 l). Other magnesium sources also work but not as well, in my experience. Putting up extra shade or applying extra paint to greenhouse roofs helps improve color under bright summer conditions. Spathiphyllums don't bother to make much chlorophyll when light levels are high. This plant usually responds well to foliar feeding, due to the somewhat soft, supple nature of the leaves.

A spray of 250 ppm giberellic acid (Pro-Gibb) successfully induces flowering in many *Spathiphyllum* varieties. It is best done with a small hand sprayer as a light, even spray to glisten, not to run off. The rate is equivalent to 0.8 ounce per gallon (23.7 ml per 3.8 l). Response time for flowering depends on variety and time of year. In general, the smaller varieties, such as Petite, bloom seven to nine weeks after spraying. The intermediate varieties, such as Tasson, take about 10 to 11 weeks. Larger varieties, such as Mauna Loa, take 12 to 13 weeks. Don't spray too early, or blooming plants may be too short. Not all varieties respond well, so test it first. Flower distortion can occur at higher rates. In addition, Benzyladenine drenched at the rate of 500 ppm successfully induces suckering in several varieties, especially the smaller ones.

INTERIOR CARE

Peace lilies tolerate the low light levels in interior environments quite well, with minimal reduction in quality. Try to provide at least 100 to 150 f.c. (1.1 to 1.6 klux), though they tolerate less for a time. Somewhat more light is even better, but the plants become pale if light is too high. Because of their jungle origins, spathiphyllums like relatively high humidity. Temperatures of 65 to 75F (18 to 24C) are fine, though for a time they can tolerate temperatures outside that range.

The plants require water fairly frequently, but don't overwater. I prefer to irrigate, wait for the plants to wilt slightly (usually in a week or so), then apply water again. Fertilize approximately every two months with about a teaspoon of 20-20-20 soluble fertilizer plus one-half teaspoon of Epsom salts. Pest problems indoors are rare with these plants, but beware of buying *Cylindrocladium*-infected spathiphyllums.

REFERENCES

Bailey, L.H., and E.Z. Bailey. 1976. *Hortus Third,* 1062. New York: Macmillan.

Blessington, T.M., and P.C. Collins. 1993. *Foliage Plants: Prolonging Quality,* 191-193. Batavia, Ill.: Ball Publishing.

Chase, A.R. 1993. Common Diseases of Spathiphyllum. *Southern Nursery Digest* (January): 20-21.

Chase, A.R, and C.A. Conover. 1988. Effect of Soil Temperature on Severity of Cylindrocladium Root and Petiole Rot of Spathiphyllum. *Foliage Digest* 11 (11): 1-2.

Chase, A.R., and R.T. Poole. 1984. Acephate Phytotoxicity of Spathiphyllum. *Nurserymen's Digest* (October): 54-55.

Chase, A.R., R.T. Poole, L.S. Osborne, and R.J. Henny. 1984. Spathiphyllum. *Foliage Digest* (July): 6-8.

Griffith, L.P. 1983. Spathiphyllum—The Peace Lily. *Florida Nurseryman* (August): 59-60.

Henny, R.J., and W.C. Fooshee. 1987. Increasing Basal Shoot Number in *Spathiphyllum* 'Tasson' With BA. *Foliage Digest* (July): 5-6.

Henny, R.J., and W.C. Fooshee. 1989. Floral Induction of *Spathiphyllum* 'Starlight' With Giberrellic Acid Treatment. *Nursery Digest* (May): 16-17.

Ott, R. 1990. Sorting Out Spathiphyllums. *Interior Landscape Industry* (March): 50-56.

Schoulties, C.L., A.R. Chase, and N.E. El-Gholl. Rev. 1983. *Root and Petiole Rot of Spathiphyllum Caused by Cylindrocladium spathiphylli.* Fla. Dept. Ag. & Consumer Svc. Plant Pathology Circular No. 218. Gainesville: Fla. Dept. Ag. & Consumer Svc., Division of Plant Industry.

SUCCULENTS:

Beaucarnea, Crassula, Euphorbia, Hoya, Sansevieria

INTRODUCTION

Succulents are generally considered to be plants with the natural ability to store water in their bodies or roots. The definition is rather loose, but most succulents are thick and fleshy, with abundant sap. In order to try to keep this orderly, each of these important succulent genera will be treated separately: *Beaucarnea, Crassula, Euphorbia, Hoya,* and *Sansevieria.* Cacti are considered members of the succulent group, but they get their own section.

BEAUCARNEA

Commonly called ponytail palms, these plants are not palms at all (not even close). The six known species are native to the drier parts of Texas and Mexico. They have been listed in both the Agave and Lily families, with the newer references indicating Agavaceae. The origin of the genus name is unknown, but the name of the most common species, *B. recurvata,* refers to the downward curvature of the leaves. Ponytails are durable and attractive as indoor plants and landscape specimens.

Production is generally from seed, with pot sizes ranging from 6 to 14 inches (15 to 35 cm). Typical peat-based potting media are generally used. Ponytails should be irrigated conservatively; allow the soil to dry out between waterings. Most growers like to use long-term, slow-release fertilizers, such as 18-6-8 or 19-6-12, on *Beaucarnea.* Another option is 20-20-20 at 200 ppm nitrogen every third watering. Ponytails are unable to respond quickly to topdressings of granular fertilizers unless they are slow-release materials.

High light is normally used, typically 4,000 to 6,000 f.c. (43.2 to 64.8 klux). The plants are heat tolerant, but it is best to keep temperature above 55F

(13C). They are rather slow growing and can be kept pot to pot during much of production. Larger specimens, especially, sell for good prices.

Problems with ponytails are relatively few. *Fusarium* causes a stem and bulb rot. Avoid excessive soil compaction and spray periodically with a thiophanate methyl fungicide, such as Cleary's 3336 or Domain. The same fungicides control *Phyllosticta,* which causes a brown fungal leaf spot. *Pythium* may cause fungal root rot when plants are kept too wet, and *Erwinia* occasionally causes a mushy, smelly soft rot. The major pests are mealybugs, mites, and scale insects. See other foliage varieties in this book for control of these various disorders.

For interior situations, high light is best, preferably 800 f.c. (8.6 klux) or better. However, ponytails can do fairly well under bright indirect light of 500 f.c. (5.4 klux) if those light levels are maintained for about 12 hours per day. Ideal temperatures are from 60 to 75F (16 to 24C), but the plants tolerate from 40 to 90F (4 to 32C) without incident. Ponytails take dry conditions rather well. They're slow growing and should be fed only about every three months. They can receive a little more feed if light levels are good.

The range between deficiency and toxicity of boron is very narrow.

CRASSULA

Jade plants have been popular as houseplants in many parts of the world. Most come from South Africa, including the common *Crassula argentea; crassula* means "thick," and *argentea* means "silvery." Commercial cultivars of *C. argentea* include Variegata, Tricolor, which has some pink in the foliage, and Dwarf Argentea.

Jade plants do best in bright light, fairly close to full sun, or a minimum of around 5,000 to 6,000 f.c. (53.8 to 64.8 klux). They must have good drainage, usually doing better with a little sand mixed into a soil based on peat and bark. Don't use too much peat, or overwatering can cause severe losses.

The main problem with jade plants is rot, caused by root suffocation from overwatering, with or without the presence of fungal pathogens. The leaves begin to shrivel, and the plant may become reddish in appearance, with soft stems. Frequently, by the time you notice rot, it is too late. The best defense is a good soil mix and conservative watering. Insect pests are generally minimal, but leaf drop can occur when exposed to 5 ppm ethylene gas. Liquid fertilization with 200 ppm nitrogen from 20-20-20 is a common feed program, occasionally supplemented with magnesium sulfate. Propagation can be accomplished any time of year, but keep leaf bud or stem cuttings in fairly low light without mist.

Fig. 40 An attractive 6-inch jade plant. (Photo by the author)

Indoors the jade plant tolerates 40 to 100F (4 to 38C), though the preferred temperature range is 55 to 75F (13 to 24C). *Crassula* tolerates as low as 75 to 100 f.c. (0.8 to 1.1 klux), though it may stretch somewhat. The preferred indoor light level is about 1,000 f.c. (10.8 klux), to maintain a jade that is reasonably compact and vigorous. It tolerates low humidity well, and the main trick is to let the soil dry out fairly well between waterings. I generally do better with clay pots for jade plants because the possibility of damage from overwatering is somewhat reduced.

EUPHORBIA

Like the jade plants, the 1,600 species of *Euphorbia* have their own family, Euphorbiaceae. The best known euphorbias are the poinsettia and the crown-of-thorns, though spurge weeds are also members of the group. They all have white, milky sap. The larger euphorbias in the foliage trade include *Euphorbia lactea,* the candelabra or milk-stripe euphorbia. It has a whitish stripe down the center of the stem and comes from the East Indies. It has never been known to flower. The African milk tree, *E. trigona,* is somewhat similar to *E. lactea,* though it usually has many more leaves near its top. Another common euphorbia is the pencil tree, *E. tirucalli,* which can grow as tall as 30 feet (9.1 m).

Euphorbias are propagated from cuttings that are allowed to harden off, or suberize, prior to planting. Production is common in full sun or in bright light of 5,000 to 6,000 f.c. (53.8 to 64.8 klux). Some nurseries maintain outdoor stock plants planted in the ground. They should not be planted near bodies of water, as the juice may be poisonous to fish. A typical foliage potting medium will do as long as it doesn't hold excessive moisture. Resist the temptation to compact the medium excessively during potting.

Problems are relatively few, but they can be severe. Whitefly can be a major pest on these plants, though since the introduction of Marathon (imidacloprid), the whitefly problem has largely been eliminated for now. During rainy periods *Phomopsis* causes unsightly, gray warts on *E. lactea* and *E. trigona*. Sprays of Daconil (chlorothalonil) have been used successfully against this problem, but Daconil is not registered on euphorbias in the United States. It is better to go with broadly labeled Cleary's 3336 (thiophanate methyl) or Dithane (mancozeb). The best control is to keep the rain exposure and overhead irrigation to a minimum. Edema, caused by wide fluctuations in moisture levels within the plant, may also develop during wet periods.

Indoors, try to keep the thorny euphorbias where people will not brush against them. They do best with some direct sun in either the morning or the afternoon. Water very sparingly, and fertilize with only about 1 teaspoon of soluble 20-20-20 per gallon (3.8 l) about every three months. If they get decent light and minimal water, euphorbias can last for years indoors.

HOYA

Commonly called wax plants, most of the cultivated types are strains of *Hoya carnosa*. The genus is named after English gardener T. Hoy, whereas the species name *carnosa* means "fleshy." They are attractive in small, 3- to 6-inch (7.5- to 15-cm) pots and in 5- to 10-inch (12.5- to 25-cm) hanging baskets. Some of the popular varieties include Variegata, with green and white foliage, as well as Tricolor, which under brighter light has green, white, and pink foliage. Another popular hanging basket variety is *Hoya argentea* cv. Picta, the Hindu rope plant. Most varieties in the trade have waxy pink and white or red and white flowers. Hoyas can bloom many times from a single flower stalk. One of my all-time favorite foliage plants is *Hoya multiflorum* Shooting Stars, whose creamy white, yellow, and black flowers resemble rocket ships.

These flowering, semiwoody vines are usually grown from 1,500 to 2,500 f.c.(16.2 to 27 klux). They do better in greenhouses than shadehouses, with steady temperatures from 68 to 75F (20 to 24C). Exposure to temperature extremes does not work. Above 90F (32C), root development is decreased. When exposed to cool temperatures, *Hoya* tends to go dormant.

Use a soluble or liquid 2-1-2 or 3-1-2 N-P_2O_5-K_2O ratio with trace elements at 200 to 300 ppm nitrogen once a week. Propagation is from single-node cuttings, usually from stock vines. The cuttings require little mist as long as humidity is 75% or above. Rooting is accomplished in three to four weeks.

Hoyas have occasional insect problems, mealybug and scale, and control measures for these insects are found in the information on other varieties in this book. Be careful, however, as both Diazinon and Orthene (acephate) are phytotoxic. I don't recommend Diazinon on *Hoya* at all. Also, while it is a rare problem, *Hoya* is nematode sensitive.

Plant germs, like people germs, are more or less everywhere, waiting for conditions to be right for infection.

For interiors, try to maintain steady temperatures of 65 to 80F (18 to 27C). Most varieties of *Hoya* do well in light levels as low as 125 f.c. (1.4 klux), though brighter light helps retain color, especially in variegated varieties. In low light, variegation decreases in the new growth, and *Hoya* generally goes partially dormant in winter. Let the soil dry somewhat between waterings, and avoid using cold irrigation water. The plants tend to bloom well if you keep them pot bound.

SANSEVIERIA

Snake plants have been in the U.S. foliage trade since the 1920s. Many Florida growers grew them in the 1930s for shipment to Europe, though today most cuttings come from the Caribbean and Central America. Other common names include "bowstring hemp" or the politically incorrect "mother-in-law's tongue." Sixty or so species exist, with the important foliage cultivars originating in drier parts of Africa and southern Asia. Most of the important types in the trade are strains of *Sansevieria trifasciata,* including the two most popular, Laurentii and Gold Hahnii. The spear sansevieria, *S. cylindrica,* is also commercially grown.

Stock plants are normally produced in open fields in Central America, where growers often use Princep (simazine) as an herbicide, though it is not labeled for such use in the United States. Propagation is from either leaf cuttings, crown divisions, harvested clumps, or rhizome division. Remember that many cultivated sansevierias are chimeras or freaks and therefore don't come true from leaf cuttings. Division is generally a faster, truer means of propagation.

Pot sizes range from 2¹/₂ to 14 inches (6 to 35 cm) for the Laurentii types. Light level varies by variety, but the variegated types usually do better in brighter light of 3,500 to 5,000 f.c. (37.7 to 53.8 klux). High rates of either dry or liquid fertilizers, at a 3-1-2 ratio, with trace elements are common. However, if temperatures below 45F (7.2C) are expected, reduce feed rates, as susceptibility to cold increases with high fertility rates. Sansevierias are damaged at 36 to 46F (2 to 8C), though the damage can take one to four weeks to show up. Cold damage usually results in large, water-soaked blotches, which look much like *Erwinia*.

Erwinia can, in fact, cause leaf and cutting rot in *Sansevieria*. Black rot of the rhizome is caused by the fungus *Aspergillus niger*. The problem is common in the Caribbean, is worse with high soil temperatures, and is controlled somewhat with thiophanate methyl fungicides, such as Cleary's 3336 or Domain. *Fusarium moniliforme* causes red to brown leaf lesions with yellow

TABLE 25 Leaf analysis rating standards for *Sansevieria trifasciata*

Nutrient (%)	Very low	Low	Medium	High	Very high
Nitrogen	<1.50	1.50–1.69	1.70–3.00	3.01–4.00	>4.00
Sulfur	<0.12	0.12–0.19	0.20–0.50	0.51–1.00	>1.00
Phosphorus	<0.10	0.10–0.14	0.15–0.40	0.41–1.00	>1.00
Potassium	<1.50	1.50–1.99	2.00–3.00	3.01–4.50	>4.50
Magnesium	<0.20	0.20–0.29	0.30–0.60	0.61–1.00	>1.00
Calcium	<0.70	0.70–0.99	1.00–2.00	2.01–3.50	>3.50
Sodium			<0.21	0.21–0.50	>0.50
(ppm)					
Iron	<30	30–49	50–300	301–1000	>1000
Aluminum			<251	251–2000	>2000
Manganese	<30	30–49	50–300	301–1000	>1000
Boron	<15	15–19	20–50	51–100	>100
Copper	<5	5–9	10–100	101–500	>500
Zinc	<20	20–24	25–200	201–1000	>1000

Sources: Institute of Food and Agricultural Sciences, Apopka, Florida; Dr. Benjamin Wolf, Fort Lauderdale, Florida.
Notes: Sample of most recent fully mature leaves, no petioles.

borders. Thiophanate methyl or Dithane (mancozeb) helps. Root knot nematodes can also be a problem.

Sansevierias have been popular indoor plants for years because of their toughness and durability under low light conditions. They can handle as little as 50 to 75 f.c. (0.5 to 0.8 klux), but they also take bright light well. Interior temperatures are best maintained between 65 and 85F (18 and 29C), and do not let the temperature drop below 50F (10C). As with most other succulents, the soil should be allowed to dry out between waterings. The best way to kill a sansevieria is either to overwater it or to expose it to chilling temperatures. A little bit of soluble fertilizer every couple of months aids in growth during the warmer months. The larger *S. laurentii* types may need to be repotted periodically.

REFERENCES

Blessington, T.M., and P.C. Collins. 1993. *Foliage Plants: Prolonging Quality,* 42-43, 79-80, 138-139, 183-184. Batavia, Ill.: Ball Publishing.

Henley, R.W. 1982. A Guide to Sansevieria Production. *Foliage Digest* (September): 3-8.

Osborne, L.S., R.W. Henley, and A.R. Chase. Wax Plant. *Foliage Digest* 9 (8): 1-4.

Rice, L.W. 1976. *Cacti and Succulents for Modern Living,* 51-53. Kalamazoo, Mich.: Merchants Publishing Company.

Stefanis, J.P., and R.W. Langhans. 1979. Commercial Production and Marketing of Succulents in Northern Climates. *Florist's Review* (November 22): 28-30, 55-60.

Wang, Y.T., and J.W. Sauls. 1988. Influence of Light, Medium, and Fertilization on Growth and Acclimatization of Ponytail Palm. *Hortscience* 23 (4): 720-721.

Watkins, J.V., and T.J. Sheehan. 1975. *Florida Landscape Plants,* 96, 98-99, 166-167, 224-227, 343. Gainesville: The University Presses of Florida.

SUCCULENTS:
Cacti

HABITAT

Cacti, the members of the family Cactaceae, are succulents. They are classified into three groups known as tribes, among which there is also a genus called *Cactus*. True cacti have areoles or warts, the nubbinlike structures from which cactus thorns emerge. About 1,500 species exist, and not all of them are spiny. The cacti evolved approximately 40 million years ago, their leaves becoming modified as spines to conserve moisture in harsh environments. Cacti are quite slow growing: A saguaro cactus 6 inches (15 cm) tall may be 10 years old. A 50-foot (15.25-m) saguaro may be 200 years old, with a root system 130 feet (40 m) across.

All cacti are native to the Western Hemisphere. Each state of the United States, except Vermont, New Hampshire, and Maine, hosts native cacti. The most important arid varieties in the foliage trade are native to drier areas of Mexico and South America. Because of the strong seasonal changes in their native environments, cacti grow and rest in cycles. They respond positively to increasing temperatures and rainfall, then tend to become dormant as the weather turns cool and dry. That generally translates into aggressive growth during the spring and

Fig. 41 *Echinocactus grusonii*, the golden barrel cactus, Homestead, Florida. (Photo by the Ed Clay)

summer months, reduced growth in fall and winter. They generally grow naturally in sandy desert areas; therefore, they prefer high light and sandy potting soils with low moisture-holding capacity.

There is a group of tree-dwelling, epiphytic cacti we call holiday cactus, which includes the Christmas cactus, Easter cactus, and Thanksgiving cactus species. These plants are native to shady, tropical jungles, especially in Brazil. They therefore prefer lower light levels in production, and while they like moisture because of their epiphytic nature, they can't stand excessively moist soil.

 Have your liquid feed analyzed, as actual injector rates frequently vary from calculated rates.

USES

Production of cactus is frequently a specialty, though foliage growers may dabble in Christmas cactus or in *Cereus peruvianus*. Pot sizes generally range from 2 to 17 inches (5 to 42.5 cm) for *Cereus* and the other large cacti. Most cacti, however, are produced in 3- and 4-inch (7.5- and 10-cm) pots. While cacti are slow growing, they require little space and relatively little care. Cacti are popular in desert dish gardens, and some imaginative planting is done using volcanic rock and driftwood. Clay or plastic pots can be used. Plastic pots are more popular with growers because of cost and durability, whereas consumers may prefer clay pots, which breathe. Plants may dry out faster in clay pots, though.

The holiday cacti are usually grown in small pots and hanging baskets for the appropriate holiday. *Zygocactus*, the Christmas cactus, can be quite durable. There is a plant in New England known to be over 80 years old. The holiday cacti do not have leaves. What appear to be leaves are actually phylloclades, modified stems.

Grafted cacti are often imported bare root from Japan and Brazil for growers to pot up and sell. Larger cacti are frequently brought in from Texas and California, though harvesting of native cacti can be severely restricted. Beware of buying cacti from guys with unmarked pickup trucks. They may be legitimate suppliers, but it is more likely that they work for Midnight Plant Acquisitions, Incorporated.

VARIETIES

The column cactus, *Cereus peruvianus,* is popular for large containers. It also comes in a spiral form. The golden barrel cactus, *Echinocactus grusonii,* is low growing and popular for medium-sized pots. There are many old man cacti, including the Peruvian old man cactus, *Epostoa lanata.* The Mexican old man cactus is *Cephalocereus senilis,* which more or less means senile head (even taxonomists sometimes have a sense of humor). Opuntias generally have flat, spiny leaves, with hairs on top of the areoles. Numerous other ornamental cacti are in the trade, as well.

The holiday cacti encompass five species and over 150 varieties. Many of them were developed at B. L. Cobia Greenhouses in Central Florida. The Easter cacti are in the genus *Rhipsalidopsis;* many of the cultivars grown today originated in Holland and Denmark. The Thanksgiving cacti are in the genus *Schlumbergera,* whereas the Christmas cacti are members of the genus *Zygocactus.*

PROPAGATION

Cacti are propagated via seeds, cuttings, offsets, and tissue culture. Varieties propagated from leaf cuttings often root best when simply laid on top of the soil, rather than pushed in. Large cacti, such as *Opuntia* and *Cereus,* are produced from trunk or large leaf cuttings, which are allowed to air-dry out of the sun for one to two weeks prior to direct sticking into the finishing container. It is best to root most cactus cuttings in bright light, but not full sun. Rooting is normally accomplished in just a few weeks.

The *Zygocactus,* or holiday cactus, is usually grown from two-leaf joints planted into dry soil. Two or three cuttings with two joints each are frequently used in 2-inch (5-cm) pots. Use three to four cuttings for a 3- to 4-inch (7.5 to 10-cm) pot, while 5- to 6-inch (12.5- to 15-cm) pots may need nine to 12 cuttings. Don't water until about a week after planting. Bottom heat to 70F (21.1C) is helpful when propagating under cool conditions.

Seeds often germinate quickly when sown in fifty-fifty mixtures of sphagnum peat and sand. Fill the pot or flat almost to the top, sprinkle the seeds, and then barely cover them. Germination time ranges from two days to one month, depending on variety and conditions. Warming the soil with bottom heat to 70 to 75F (21.1 to 23.9C) hastens germination. Most varieties need light to germinate, except the genus *Parodia.* It normally takes nine to 12 months to finish a cactus from seed.

Grafting of desert cacti is also practiced, often between different species. Use a very sharp, sterile knife. Popular graft methods are the cleft graft, which is much like a tongue-and-groove joint. Flat and side grafting are also used. It is important to line up the vascular bundles of the two pieces being grafted and secure the union with rubber bands, tape, or even cactus thorns.

CULTURE

It is helpful to wear heavy gloves when handling cacti, and many growers use tongs, baskets, and specialized tools for cactus production. Most arid cacti prefer bright light close to full sun; 5,000 f.c. (53.8 klux) is desirable for seedlings, 8,000 f.c. (86 klux) for mature plants. When cacti are actively growing, temperatures can range up to 95F (35C), while dormant cacti prefer a rather cool 45 to 55F (7 to 10C).

It is imperative that soil moisture-holding capacity be low for cactus production. Use a forgiving mix relatively low in organic matter and with a high percentage of sand. Popular growing media include typical potting soil blended with one-half to two-thirds by volume of coarse sand. Mixtures of peat, sand, and perlite are also used. Volcanic gravel is a nice addition, if available. Some growers like to mix in pieces of charcoal to help maintain high pH.

The general rule on irrigating cacti is to water well but infrequently. Cacti do need water, especially in spring and summer. It helps to feel the moisture level in the soil and sense the container weight to determine whether irrigation is required. Wet the soil thoroughly, then allow it to dry. Dormant cacti in winter usually need to be watered every few weeks, just enough to avoid shriveling.

The holiday cacti, which are jungle epiphytes, prefer a peaty mix, though with 20 to 25% sand by volume. Some growers use a mixture of 40% perlite and 60% peat, limed to a pH of 5.5. Preferred light levels for holiday cacti are 65 to 80% shade (1,500 to 3,000 f.c., or 16.1 to 32.2 klux). Best temperature ranges are 70 to 85F (21.1 to 29.4C), though the plants will take 40F (4.4C). In winter it is best to maintain a minimum greenhouse temperature of 50 to 60F (10 to 15.6C). Try to keep a 60F (15.6C) minimum soil temperature for *Zygocactus*. During production it helps to pinch off the leggy-growing tips in a procedure known as leveling. Holiday cacti generally require eight to 12 months for finishing.

NUTRITION

Being slow growers, cacti have relatively low fertility requirements. Low-nitrogen, high-phosphate fertilizers, such as 10-20-10 or 10-20-20, are generally preferred. Many soluble African violet fertilizers are acceptable for cactus production. Slow-release, coated fertilizers at 18-6-12 or 18-6-8 are sometimes used for the larger cactus varieties, though the formulation should be long term and the rates conservative. It does help to have trace elements incorporated into the potting mix for most cacti.

Zygocactus is usually grown with soluble fertilizer, often 20-20-20 at 150-200 ppm nitrogen, one to two times per week. It helps to leach about two times per month. Trace element problems and deficiency symptoms are not very common in cacti, though holiday cacti may have problems with lack of magnesium or iron. Magnesium deficiency results in a yellowing or reddening of the older growth, whereas iron deficiency causes veinal chlorosis in newer foliage. Be careful, however, as chlorosis from cold soil temperatures looks very much like iron chlorosis. Drenching with chelated iron usually alleviates iron chlorosis in *Zygocactus*.

DISEASES

The most common fungal disease of cacti is *Drechslera cactivora*, which often causes stem rot. The onset of the disease can occur very quickly, perhaps only two to four days. The basal portion of the plant begins to turn yellow or dark green, generally becoming dark brown. Many cacti suffer from this disease, including *Cereus* and the barrel cacti. Wounding of tissue helps the fungus gain entrance. In *Cereus* cacti the disease may attack upper portions of the plant, entering through the stomata. Sprays of Chipco 26019 (iprodione) or Carbamate (ferbam) are helpful, and these are labeled for cacti. Some growers use Captan or Daconil (chlorothalonil), but these fungicides are not registered for cacti in the United States.

While *Drechslera* causes a dry basal rot, wet basal rot may be caused by either *Pythium debaryanum* or *Phytophthora cactorum*. In both cases the symptoms will be dark brown to black, wet-looking basal tissue. Immediately spray with Aliette (fosetyl-aluminum) or drench with Subdue (metalaxyl), then dry the plants down. *Fusarium* and several other fungi may cause rots in cacti, though they are somewhat less common.

Erwinia causes bacterial rot of cacti under wet, humid conditions. The affected tissue is very mushy and foul smelling. Little can be done by the grower at this point, though sprays of Phyton 27 (picro cupric ammonium formate) may help prevent spread. An old-fashioned technique is to cut out diseased portions of cacti with a sharp grapefruit spoon, then dust the wound with powdered sulfur.

INSECT AND MITE PESTS

Mealybugs and root mealybugs are occasional problems for cactus growers. The white, cottony masses may be observed on aboveground parts or on roots. As the infestation increases, plants lose vigor and become discolored. Sprays with Malathion or Cygon (dimethoate) are helpful. Several scale insects attack cacti from time to time. Treatments are similar to those for mealybugs.

Once in a while spider mites attack a few cactus varieties, generally those with parts that are less succulent. Insecticidal soaps are helpful, as are Pentac (dienochlor), Avid (abamectin), and others. A moth called *Cactoblastis* has destructive larvae that are a significant problem for opuntia growers in Florida and the Caribbean. I have observed severe infestations on the island of Hispaniola. In fact, this insect has been used to control wild populations of opuntia cacti with considerable success. Systemic insecticides, such as Cygon (dimethoate) or Orthene (acephate), may provide some control.

DISORDERS

The most common disorder of cacti is rot, usually caused by excessive watering, which may result in either root suffocation or fungal infection. Plants shrivel if severely dry, though root rot can also cause shriveling. A curious disorder known as cresting, also as fasciation, can happen with *Opuntia, Cereus,* and *Echinopsis* cacti. The plant develops not one growing point, but many growing points, resulting in a very curled, gnarled appearance. Insect activity and radioactive soils have been known to induce fasciation. Some of the bizarre crested types are desirable as ornamentals, however.

When seedlings turn reddish, it is generally because of too much light. Too little light causes seedlings to be pale green, and they may stretch. Low soil temperatures result in a chlorosis which looks very much like iron chlorosis. Sunburned tissue may result in substantial amounts of yellow or white tissue on the foliage. Edema occurs when cacti are kept dry in hot conditions, then suddenly cooled off or watered heavily. The excess moisture exudes through the intercellular spaces, resulting in nonpathogenic bumps called edema. Good moisture management is the only control.

Holiday cacti are very sensitive to ethylene, generated by improperly burning heaters, tractor exhaust, rotting fruit, or waterlogged soils. Massive flower drop occurs when plants are exposed to ethylene. The older leaves may become yellowed and shriveled, and these leaves will ultimately drop.

Zygocactus may turn a dull blue to gray color when kept too wet or if soil is poorly aerated. Rot often ensues under such conditions. Dry the plants out when this symptom is observed. Unfortunately, the holiday cactus may also turn blue to gray when excessively dry. Hard water may cause staining of cactus foliage, in which case water treatment systems or subirrigation systems are useful.

TRICKS

Don't water cacti right away after planting or transplanting. Let any wounded tissue and roots heal and suberize. Wait about a week before the first watering. A little charcoal in the potting mix helps keep soil pH toward the high side. Many cacti respond well to the calcium and phosphorus in bonemeal incorporated into the mix. Mature cacti may be top-dressed with bonemeal, especially in the spring.

Barrel cactus always leans a little bit to the south. You can tell *Fusarium* rot in cactus from other rots, in that *Fusarium* rots generally have a yellow margin on the edge of the infected area.

Because cacti are desert plants, it is safe for their seeds to germinate and establish themselves only during rainy periods. Otherwise, the seedlings die quickly. Cacti therefore have germination inhibitors, which must be leached away for germination to occur. Use a mix that is very well drained, low in moisture-holding capacity, and sterile. Soak the seeds thoroughly at planting to leach away the inhibitors, then maintain adequate moisture early in propagation.

Wear heavy gloves or use sections of newspaper when handling cacti. Better yet, get someone else to do it. Sometimes after handling the plants, you will feel the glochids--tiny, barbed hairs which feel like fiberglass threads--on your skin. Apply rubber cement or cellophane tape to the affected area, and the glochids should pull off fairly easily.

INTERIOR CARE

Cacti perform well in artificially heated and cooled environments, where humidity is often low. Give them as much light as you can and, again, use a very light, nonabsorbent soil mix. Don't be afraid to water cacti during the warmer parts of the year. In winter, though, irrigate only every few weeks to prevent shrivel. Don't fertilize dormant cacti or unrooted cuttings. Begin fertilizing only in spring, continuing through summer, using a light rate of 20-20-20 or a similar soluble fertilizer.

Repotting is best done during March and April, prior to the active growing season. Good lower-light cacti include the Haworthias, also known as zebra cacti. The holiday cacti tolerate low light fairly well, but they bloom and retain their flowers much better under bright or artificial indoor lighting.

Phytophthora means "plant destroyer" in Latin.

REFERENCES

Boyle, T.H., and D. Stimart. 1989. A Grower's Guide to Commercial Production of Easter Cactus. *GrowerTalks* (November): 50-52.

Boyle, T.H. 1992. Commerical Production of Easter Cactus. *Foliage Digest* (May): 3-6.

Habeck, D.H., and F.D. Bennett. 1990. Cactoblastis cactorum berg, *a Phyticine New to Florida*. Entomology Circular No. 133: 1-3. Gainesville: Florida Department of Agriculture and Consumer Services, Division of Plant Industry.

Klenn, J. 1984. A Florida Cactus Family. *Florist's Review* (June 21): 26-29.

Lamb, E., and B. Lamb. 1970. *The Pocket Encyclopedia of Cacti in Color*. New York: Macmillan.

Manning, R. 1941. *What Kinda Cactus Izzat?* 27-33. New York: J.J. Augustin Publisher.

Rice, L.W. 1976. *Cacti & Succulents for Modern Living*, 2-11. Kalamazoo, Mich.: Merchants Publishing Company.

Ridings, W.H. 1972. *A Stem Rot of Cacti*. FDACS Plant Pathology Circular 191: 9. Gainesville: Florida Department of Agriculture and Consumer Services, Division of Plant Industry.

Rowley, G.D. 1978. *The Illustrated Encyclopedia of Succulents and Cacti*, 92. New York: Crown Publishers, Inc.

Scott, S.H. 1958. *The Observer's Book of Cacti and Other Succulents*, 13-31. New York: Frederick Warne & Co.

Slade, I. 1982. Zygo Christmas Cacti. *Florist's Review* (September 16): 26-29.

SWEDISH IVY

(See Other Foliage: Plectranthus)

SYNGONIUM

Nephthytis

HABITAT

Syngonium, commonly known as the arrowhead vine or nephthytis, has been popular as a houseplant all over the world for many years. The genus name *Syngonium* refers to cohesion of the plant ovaries in Greek. Most cultivated varieties today come from *Syngonium podophyllum* (foot-leaf), which is native from Mexico to Panama. There are about 20 other species, most of which are native to Central America. The old genus name of *Nephthytis* still remains as a common name, though botanically *Nephthytis* today refers to four species

Fig. 42 *Syngonium* White Butterfly in an 8-inch pot, Jamaica. (Photo by the author)

of African herbs distinctly different from syngoniums. True *Nephthytis* is probably not in the ornamental trade at all.

Syngoniums are generally found in Central American jungles, growing as ground covers and sometimes up and over trees and rocks, similar to philodendrons. They may at times be found at significant elevation, hence their mild degree of cold tolerance.

USES

Popular with foliage growers due to their fast crop time, syngoniums have a wide array of potential uses. Hanging baskets from 5 to 10 inches (12.5 to 25 cm) are produced, as are potted specimens from 3 to 8 inches (7.5 to 20 cm). Dish gardens and terrariums almost always contain at least one syngonium because its short, bushy habit and tropical appearance are well adapted to that kind of culture. Mass plantings, ground covers, and hanging baskets are the most common interior applications. Eight- and 10-inch (20- and 25-cm) hanging baskets are very popular for restaurants and lounges.

VARIETIES

The cultivar White Butterfly has been the most popular variety in the U.S. trade since the early 1980s. The plant has arrow-shaped leaves of a somewhat dull, emerald green color, with mottled white pigmentation on the upper leaf surfaces. The plant is rather upright, self-heading, and bushy. Pink Allusion is similar to White Butterfly, though it has pink veins in the center of the leaf. Bob Allusion, which originated at Donaldson's in Zellwood, Florida, has stronger pink veins than Pink Allusion and is popular for dish garden production. Cream is a popular small variety, its cream-colored variegation contrasting with the green foliage. Lemon-lime has a variegated pattern of yellow and pale green. Pixie, a dwarf variety, is popular for dish gardens and 3-inch production. Lemon-lime and Pixie were selected and named by Mark Poorbaugh of Prolific Plants, Apopka, Florida.

PROPAGATION

Most *Syngonium* propagation today is from tissue culture. This plant lends itself well to tissue-culture production, which has enabled large quantities of new varieties to reach the trade quickly. The compact, clumping nature of tissue-culture plants also helps yield compact, uniform, disease-free starter plants.

Prior to tissue culture, syngoniums were rooted from leaf-and-eye cuttings. Like most other aroids, they root rather easily, as long as only minimal mist is applied to avoid disease problems.

CULTURE

Syngoniums are among the most heat-tolerant foliage plants. They do fine at 105F (40.5C) and tolerate temperatures ranging from 63 to 106F (17.2 to 41C). The ideal temperature range, however, is from 70 to 95F (21.1 to 35C). Growers in more northern, cooler climates can successfully produce syngoniums as long as they maintain both air and soil temperatures between 60 to 70F (15.6 to 21.1 C). It is preferable to grow these plants in a greenhouse under cover, as exposure to seasonal rainfall in a shadehouse can create substantial disease problems. The plants prefer a steady moisture supply and reasonably high humidity.

Typical light levels are 1,500 to 3,500 f.c. (16.2 to 37.7 klux). This is roughly equivalent to 60 to 80% shade. In brighter light, such as 47% shade, the plants send more but smaller leaves. The specific light requirements depend a little on the cultivar, as both light and temperature can affect color, plant quality, and variegation. You may need to experiment to find the best combination for your variety. Potting media are usually composed primarily of sphagnum peat, typically around 40 to 50%, with the remainder coming from composted pine bark, perlite, or polystyrene beads. The mix is usually limed to a pH of around 6.0.

NUTRITION

As opposed to many foliage plants, syngoniums are known to tolerate a wide variety of fertility rates. The ideal fertilizer level depends in part on the light level, but these plants seem to grow reasonably uniformly from very low to very high rates of feed. A 3-1-2 ratio is typical, and many growers use coated, slow-release 19-6-12 or 18-6-8 at an average rate of 12 grams per 6-inch (15-cm) pot every three months. The actual rate would depend on the fertilizer formulation. Liquid feed from 9-3-6 or 20-10-20 at 200 ppm nitrogen is most common for smaller pots. Nitrogen source appears to be relatively unimportant, though growers in the far north may do better with more nitrate nitrogen in the wintertime. It doesn't seem to matter in the tropics.

Deficiency symptoms are not very common when growing this plant. Lack of nitrogen will result in small, pale plants. Some of the greener varieties display the normal marginal yellowing of older leaves when magnesium is deficient. This generally happens only with rather mature specimens. Manipulation of light and magnesium levels can be important in achieving the right amount of color contrast in variegated varieties. In order to stimulate early root growth when starting plants off, it helps to incorporate triple super phosphate into the mix or to include a high-phosphate starter fertilizer in your initial fungicide drench. Low calcium causes weak, thin stems and leaves.

DISEASE

Three bacterial diseases primarily affect syngoniums. *Erwinia* causes an irregular, mushy, water-soaked leaf spot, which ultimately turns tan to dark brown. Unfortunately, syngoniums also have their own strain of *Xanthomonas, X. campestris* pv. *syngonii*. This bacterium generally enters through hydathodes, which are moisture-excreting glands in the leaf margins. Therefore, *Xanthomonas* is found more on the edges of the leaves than *Erwinia,* and the lesions appear clear to yellowish and are also water soaked. *Xanthomonas* can kill a syngonium under the right conditions. *Pseudomonas cichorii* also causes smaller bacterial leaf spots, especially when the foliage stays wet. To control bacterial leaf spots on this plant, keep the leaves dry, use clean stock, and practice good sanitation. Chemicals are of only limited benefit, but the registered ones include Agrimycin 17 (streptomycin sulfate), Phyton 27 (picro cupric ammonium formate), and Kocide 101 (cupric hydroxide).

Probably the most common fungal disease is *Myrothecium,* which frequently attacks young tissue-culture plants. The fungus causes a leaf and petiole rot, with small, greasy-looking lesions, usually with black and white fruiting bodies visible on the edges of the spots. Don't fertilize too heavily, or *Myrothecium* will be more severe. Sprays of Chipco 26019 (iprodione) or Terraguard (triflumizole) should control it. *Cephalosporium* is a fairly uncommon foliage disease, but it does cause fungal leaf spot of *Syngonium,* especially in rainy weather. Small, roundish, reddish brown spots form with a yellow halo. Sprays of Daconil (chlorothalonil) or Dithane (mancozeb) are effective.

Ceratocystis, another uncommon disease of foliage plants, causes black cane blight and black stem canker in this plant. It can also cause leaf spot or root rot symptoms. Fungicides are generally not effective. Captan has been used, though it is not labeled for this plant in the United States. Hot-water dips of 30-minute duration in 120F (49C) water help rid cuttings of the fungus. Aerial *Rhizoctonia* frequently causes brown lesions of the lower leaves, with visible reddish brown threads where leaves touch the soil. Controls are the same as for *Myrothecium.*

INSECT AND MITE PESTS

In contrast to *Syngonium* disease problems, insect pressure is rather mild. Mealybugs display their usual white, cottony masses on leaf axils, lower leaf surfaces, and roots. Sooty mold may form in the juice secreted by the insects. Discard badly infested plants, then spray the remainder with Dycarb (bendiocarb) or Mavrik (tau-Fluvalinate). Diazinon drenches may be helpful for root mealybugs. Scale infestations are rather rare in this plant. Controls are the same as for mealybugs.

Spider mites, especially red mites, attack at times. Predator mites, such as *Phytoseiulus,* are beneficial, as are sprays of insecticidal soaps. Kelthane (dicofol), Pentac (dienoclor), or Avid (abamectin) also controls them, but angling the spray to the underside of the leaf can be difficult. Whiteflies cause problems at times. Granular Marathon (imidacloprid) works well, as do several other materials. Thrips are sometimes a problem, causing leaf distortion and curl, with silver-gray scars visible on leaves and petioles. Sprays of Mavrik (tau-Fluvalinate), Orthene (acephate), or Avid (abamectin) should help.

DISORDERS

Syngoniums like a steady supply of moisture, and when they run dry, the older leaves tend to turn solid yellow and then drop. In winter, young leaves may display a disorder of water-soaked leaves commonly called dead spots. It happens when soil is cold and air is warm, causing a moisture imbalance and moisture loss in young leaves. Cold irrigation water and cold condensate from greenhouse roofs worsen the disorder. Try to maintain the soil temperature at 65F (18.3C) or above, and after cool nights slowly raise greenhouse temperatures the next morning.

Under low light levels these plants become rather stretched and weak. Do what you can to improve light levels; calcium sprays may strengthen the plants a little bit. Copper toxicity is not common, but it can happen in syngoniums. It is hard to distinguish it from some of the diseases listed previously. The leaf spots from copper toxicity, however, tend to be rather dry and brown and uniformly distributed throughout the group of plants.

TRICKS

The most important *Syngonium* cultural trick is to maintain a steady supply of moisture, at the same time irrigating so that the foliage stays wet only a minimal amount of time. It doesn't like cold irrigation water, so don't water early in the morning, when the water is cold (outside of the tropics). Try to irrigate in the middle of the day. Otherwise, either fill a tank in the greenhouse, where the water will warm up, or use a water heater.

Increasing fertility rates will reduce incidence of *Xanthomonas* but does not affect *Erwinia* either way. When *Erwinia* lesions are more tan in color, especially during dry weather, it indicates that the bacterium is relatively inactive. Darker lesions indicate *Erwinia* that is more active.

Beware of copper sprays when used in the rotation with such acidic materials as Aliette or vinegar. The combination increases the chances of copper toxicity. Increasing the feed rates intensifies the white color of White Butterfly, but going too high may reduce plant quality. Pink Allusion should be grown somewhat darker than other cultivars, in the range of 1,000 to 1,500 f.c. (10.8 to 16.1 klux). It is best to spray for *Myrothecium* when temperatures are between 60 to 80F (15.6 to 26.7C), when the fungus is most active.

 Yellow sticky traps are useful for capturing whiteflies and fungus gnat larvae.

INTERIOR CARE

Syngoniums are durable, forgiving houseplants for most situations. Try to maintain temperatures between 65 and 85F (18.3 and 29.4C). Bright, indirect light is best. The minimum light level for decent quality is 75 f.c. (0.8 klux), though a better working minimum light range would be 100 to 150 f.c. (1.1 to 1.6 klux). Variegation generally decreases in most varieties under low light.

Check for signs of mealybugs and mites from time to time. Try to keep soil moisture levels fairly consistent and maintain at least 40% relative humidity. If light levels are good, you can fertilize every two months with 1 teaspoon of soluble 20-20-20 per gallon. In low light it is best to keep fertility low, or the plants may develop weak growth with poor color. As long as syngoniums receive reasonable care, they can be maintained for many years in the interior environment.

REFERENCES

Blessington, T.M., and P.C. Collins. 1993. *Foliage Plants: Prolonging Quality,* 194-196. Batavia, Ill.: Ball Publishing.

Chase, A.R. 1987. *Effect of Fertilizer Rate on Susceptibility of Syngonium podophyllum 'White Butterfly' to Erwinia chrysanthemi or Xanthomonas campestris.* CFREC-A Research Report RH-87-5. Apopka: Central Florida Research and Education Center.

Chase, A.R. 1988. *Effect of Nitrate-Ammonium Ratio on Growth of Syngonium podophyllum 'White Butterfly' and Susceptibility to Xanthomonas campestris* pv. *Syngonii.* CFREC-A Research Report RH-88-12. Apopka: Central Florida Research and Education Center.

Chase, A.R. 1994. What Ails It? Common Diseases and Disorders of Syngonium. *Landscape and Nursery Digest* (December): 33-35.

Chase, A.R., L.S. Osborne, and R.T. Poole. 1984. Syngonium. AREC-A Foliage Plant Research Note RH-1984-G: 1-6. Apopka: Agricultural Research and Education Center.

Chase, A.R., and R.T. Poole. 1985. *Effects of Temperature on Growth of Syngonium 'White Butterfly'.* AREC-A Research Report RD-85-20: 1-4. Apopka: Agricultural Research and Education Center.

Chase, A.R., and R.T. Poole. 1988. Fertilizer, Temperature and Light Affect Growth of *Syngonium* 'White Butterfly'. *Foliage Digest* 11 (7): 1-4.

Foliage Plants: New Varieties Go Beyond Green. 1990. *Greenhouse Manager* (February): 18.

Poole, R.T., and C.A. Conover. 1990. Growth of *Cissus, Dracaena* and *Syngonium* at Different Fertilizer Levels, Irrigation Frequencies, and Temperatures. *Foliage Digest* 13 (8): 1-3.

Poole, R.T., and C.A. Conover. 1993. Light Intensity and Fertilizer Rate Affect Growth of *Syngonium podophyllum* 'Pink Allusion'. *Southern Nursery Digest* (March): 23-24.

Poorbaugh, M. March 1997. Telephone conversation with the author.

Welker, R. March 1997. Telephone conversation with the author.

UMBRELLA TREE

(*See* Schefflera actinophylla)

WANDERING JEW

(*See Other Foliage:* Zebrina)

WARNECKII

(*See* Dracaena deremensis)

WEEPING FIGS

(See Ficus benjamina *and* Ficus retusa*)*

YUCCA

HABITAT

Most yuccas are native to the southern United States and Central America. Approximately 40 species exist, the most common one in the trade being *Yucca elephantipes,* the spineless yucca. The name *Yucca* comes from a modification of an ancient aboriginal name. *Elephantipes* means "elephant foot." In mature specimens the thick, basal portion of the yucca trunk resembles the foot of an elephant. In the summer, the plant has white, lilylike flowers that are edible. Flowers are rarely seen in foliage production, however.

Yucca elephantipes is native to Mexico, though it is most widely cultivated in Guatemala.. I have seen a number of mature specimens that easily reach 30 feet (9.1 m) tall. The yucca has a thick trunk and thick branches with numerous heads emanating from the middle and upper portions. It actually grows more as a tree in its native habitat. The rigid, dull green leaves maintain good color even in full sun. In Central America yuccas grow in tropical pastures and forests, including the higher elevations. I have also seen yuccas fairly close to the sea on the Pacific side. Spineless yucca is generally said to grow in USDA Zones 10B to 11.

USES

Spineless yucca tips are commonly rooted as single plants per 6- to 10-inch (15- to 25-cm) container. Larger containers are occasionally planted with three tip cuttings each. Standards are sometimes produced by growing a single cutting to a height of about 2 feet (0.6 m), then cutting off the tip. The resulting multiple head combines with the thick, original trunk to form a standard type of plant. Large, branched stumps are occasionally imported and rooted in larger containers.

The most common usage, however, is to plant two to four canes of staggered heights directly in a 10-inch (25-cm) or 14-inch (35-cm) container. The heads will sprout from the various heights, making an attractive woody foliage plant.

Yucca cane was popularized by Lex Ritter in the

Fig. 43 Yucca cane from Homestead, Florida, showing multiple sprouts. (Courtesy of Ed Clay)

mid-1970s. He and his son collected yucca cane from Nicaraguan coffee plantations and from the Guanacaste region of Costa Rica and shipped them to the United States. Yuccas are widely used in landscapes in the tropics, and it is common to find them on the edges of coffee plantations as living fences.

They may call it spineless yucca, but I once poked out an eardrum bumping into a leaf tip while inspecting a block of yucca cane.

VARIETIES

The majority of production involves just the one species, *Yucca elephantipes*. However, you will occasionally see *Variegata*. This plant is a surprisingly strong grower, frequently more vigorous than the species. It is rare to have a variegated cultivar outperform its green parent, but in biology there are always exceptions. *Variegata* has a white to yellow stripe along the edge of each leaf, just inside the margin. Culture is basically identical to that of the species, though the variegated type finishes a little faster.

Don't spray foliage plants when the temperature exceeds 85F (29.4C).

PROPAGATION

It is possible to propagate yuccas by seeds, offsets, cuttings, cane, rhizomes, and root pieces. Commercially, only tip cuttings and woody cane pieces are propagated. Yucca tip cuttings come in various sizes, depending on the intended use. Most growers remove three or four of the oldest leaves prior to inserting the cuttings, with the base going 2 to 4 inches (5 to 10 cm) into the soil. Mist is not required for cutting propagation, but do maintain reasonable soil moisture. A high rooting percentage is normally accomplished in two to three weeks under warm conditions. Cuttings can also be rooted in full sun.

Some cane is still collected from the wild by natives in the rural highlands of Central America, but most cane comes from yucca stock farms where long pieces of cane from the field are cut and brought to a processing and packing area. There the cane is sorted by stem caliper size and cut to desired lengths. When cutting from stock plants, avoid young and skinny canes as well as old and fat ones. The sorted cane pieces are then usually dipped on either end in a fungicide, and the top end is sealed by dipping in melted paraffin wax. Cane is sometimes shipped with the pieces upright and crated, and at other times it is bundled and shipped horizontally.

Cane should be planted promptly after it is received. It may be rooted in beds, but most growers plant directly into finishing containers. Put a few inches of soil into the bottom of the container, orient the cane pieces and fill with more soil. Irrigate briefly once a day just to maintain adequate moisture.

Cane pieces generally root in about a month, depending on size and time of year. Fresh cane should root promptly and at high percentages, whereas older cane rooting may be somewhat erratic. Old cane frequently has an elliptical hollow spot on the end where the tissue has dried and contracted. Avoid such canes if possible, or at least recut the ends.

CULTURE

Many growers prefer to start and grow yuccas in full sun. Canes break well in full sun, and the tip cuttings will form strong, sturdy plants. Frequently, after the plants are a few weeks from marketable size, they are moved into a shadehouse with 50 to 60% shade, or around 5,000 f.c. (53.8 klux). You can acclimate yucca cane in less light, though the plants can stretch and become weak if you leave them too long. The minimum useful light level is probably 3,000 f.c. (32.3 klux).

Optimum temperatures are 65 to 95F (18.3 to 35C), though temperatures outside that range are tolerated fairly well. Cold injury tends to occur around 40F (4.4C). Most potting media for yucca are peat based, usually with some bark and up to 20% coarse sand for weight and support. The plant needs fairly regular irrigation. Even though it looks somewhat cactuslike, yucca is not a cactus. It is generally considered to be in the Agave family, but some list it in the lily family.

NUTRITION

Most yucca producers use granular fertilizer of approximately a 3-1-2 ratio of N-P_2O_5-K_2O, with magnesium and trace elements. It doesn't hurt to incorporate minor elements into the mix, though it is not essential, either, as trace element deficiencies are not very common in yuccas. They are somewhat tolerant of soluble salts, though excessive fertilizer can give them severe, brown scorching of the older foliage. It has not been documented, to my knowledge, but I believe yuccas are more sensitive to ammonia than to soluble salts, so try to avoid high-ammonia fertilizers. When producing smaller cuttings from tips, it may be more practical to use liquid fertilizer, either 20-10-20 or 24-8-16, at approximately 200 ppm constant liquid feed, with a leach every two weeks.

Nutrient deficiency and toxicity symptoms are fairly rare in spineless yucca. Lack of nitrogen appears as pale green to yellow plants with narrow, strappy leaves and little growth. About the only nutrient deficiency commonly observed is lack of boron. Low boron causes yucca leaves to bend downward in the middle, giving a somewhat droopy effect. This is more common in low light situations, but it can also happen in full sun. A foliar spray with Solubor or chelated boron generally results in a good response. Severe boron deficiency causes acute deformity in the emerging leaves, and the heads from canes fail to develop.

DISEASES

The most troublesome disease of spineless yucca is *Coniothyrium concentricum*, sometimes called shotgun fungus. Tiny, clear spots develop, primarily on the older leaves. The spots are circular and about the size of BBs. They turn brown to black with maturity and occasionally join to form larger blotches. Control measures include keeping foliage dry, irrigating when drying conditions are good, and spraying with Daconil (chlorothalonil) or Dithane (mancozeb). Removal of older infected leaves is helpful, at least in the early stages of infection. *Cytosporina* is less common, and it causes dark lesions, especially on the leaf margin near the leaf tip. It is brown to black and usually affects only one or two leaves per plant. Sprays of Daconil generally control it.

Both *Fusarium* and *Erwinia* can be found causing cane rot in yuccas, usually within the first few weeks after planting the cane. *Erwinia* creates a wet, mushy, smelly rot, whereas with *Fusarium* the bark loosens from the internal woody tissue, and the inside is dry. *Erwinia* is very difficult to control with chemicals, though Phyton 27 (picro cupric ammonium formate) may be of some benefit. *Fusarium* is usually controlled with a thiophanate methyl fungicide, or Captan. Decent soil moisture, healthy cane, and good drainage are the primary ways to avoid these diseases. Southern blight occasionally causes stem rot, with white mycelia growing on the bark and the soil surface. I have never encountered this disease on yuccas, but a spray and drench with Terraclor (PCNB) should control any occurrence.

INSECT AND MITE PESTS

A serious insect pest is the yucca weevil, *Scyphophorus acupunctatus*. This black weevil has a strong, sharp snout, hence the species name. It bores into the base of cane pieces, and the eggs hatch to form grublike larvae. These tunnel through the interior portions of the cane, causing the plant to collapse, topple over, and ultimately die. The insect is native to the southwestern United States and Central America and has been known since 1840. I have seen severe infestations in El Salvador. Sprays of Cygon (dimethoate) at two-week intervals may be helpful. Granular Di-syston (disulfoton) may also help. Lindane and Dursban (chloropyrifos) are two other options.

Thrips feeding causes numerous white to silver-gray scars, which appear on the newer leaves and the emerging spear. Sprays of Orthene in combination with a wetting agent should control them. You need the wetting agent to help the spray run down into the whorl, where the thrips frequently hide. Scale and mealybug infestations are occasionally seen in yuccas, especially on plants propagated from tip cuttings. Sprays of Dycarb (bendiocarb) or Cygon (dimethoate) usually control these insects.

Ambrosia beetles (*Xyleborus*) sometimes attack newly planted yucca cane that is under stress. Small, thumbtack-size holes can be seen, frequently with sawdust emerging from them. The insect looks like a tiny, brown cockroach.

Bad cane pieces should be discarded, and the remainder treated with Lindane. Central American cane is frequently bothered by an insect called chinche (genus unknown), which causes rasping injury to the foliage.

DISORDERS

Yuccas are somewhat fluoride sensitive. Accumulation of fluoride from superphosphate fertilizers or high-fluoride irrigation water tends to cause tip and marginal burn. It is not a severe problem for the grower, as usually the few bad leaves can simply be removed. A sudden increase in light levels tends to cause irregularly shaped, blanched areas on the upper leaf surfaces. They usually occur toward the center of horizontally oriented leaves, sometimes on the younger, more vertical leaves. Try to increase light levels in yuccas only gradually, as sunburn can be a significant problem. Sometimes this disorder can be observed on canes and cuttings, grown in the sun, that fail to root.

Another common disorder is head breakdown, where yucca canes sprout heads that may grow a few inches, then begin to die back. Heads become pale green and withered looking, with tip burn and numerous water-soaked lesions in the leaves. This also is caused by a simple failure of the cane piece to root. The head comes out, and the roots intended to support them fail to develop. Discard cane pieces that show these symptoms.

TRICKS

Maintaining high fertility levels goes a long way toward controlling *Coniothyrium* leaf spot. The plants are much more susceptible when low in nitrogen. The leaf spot takes 21 days to appear from the time it is actually initiated. You may have received apparently clean tip cuttings, but have the spots become visible shortly thereafter. You should begin Daconil (chlorothalonil) sprays right away when planting tip cuttings.

Spineless yucca is somewhat salt tolerant, though less so than many of its fellow species. Be very careful about applying preemergent herbicides to newly planted yucca tips, as the chemical can accumulate in the whorl and cause severe damage and even cutting death.

If too many heads develop from cane pieces, thin them back to about four well-distributed heads. If heads come out eight or 12 at a time, they will fail to develop adequate size. If only a single head sprouts, break that off quickly, and several more will usually sprout behind it, creating a fuller effect. If heads are not developing well, spray with boron. It also helps to orient multiple cane pieces with shorter canes facing an aisle so they will receive more light.

INTERIOR CARE

Yucca tips and canes do fairly well indoors if they can receive at least 150 f.c. (1.6 klux) of light, though 250 f.c. (2.7 klux) is better. Irrigate the plants fairly thoroughly, then permit them to dry down before watering again. Try to use low-fluoride irrigation water; avoid fluoridated city water, or tip burn problems may develop. Generally, it is not a good idea to fertilize yuccas indoors because the combination of fertility and low light creates rather weak, floppy, unattractive growth. The plants generally maintain color just from the nutrients generated by media decomposition. Don't use leaf shine products on yuccas, as they can be injurious, and yucca leaves don't shine anyway.

REFERENCES

Bailey, L.H., and E.Z. Bailey. 1976. *Hortus Third*, 1178-1179. New York: Macmillan.

Poole, R.T., L.S. Osborne, and A.R. Chase. 1985. *Yucca*. AREC-A Foliage Plant Research Note RH-1985-J: 1-5. Apopka: Agricultural Research and Education Center.

Pott, J.N. 1975. A Yucca Borer, Scyphophorus acupunctatus, in Florida. *Fla. State Hort. Soc.* 88: 414-416.

Ritter, A. March 1997. Telephone conversation with the author.

Watkins, J.V., and T.J. Sheehan. 1975. *Florida Landscape Plants*, 107. The University Presses of Florida.

ZEBRA PLANT

(*See* Aphelandra)

OTHER FOLIAGE VARIETIES

This section contains a brief coverage of 17 other foliage varieties that are to some degree specialty items. They are relatively minor varieties compared to those that received a full review, but they are still grown in significant numbers.

ACALYPHA

Two species of *Acalypha* are occasionally grown in larger containers as foliage plants. The copperleaf, *A. wilkesiana,* has variegated foliage colored off-white and copper. The chenille plant, *A. hispida,* has coarse, deep green leaves with fuzzy, drooping, bright red flowers. I have seen them called dreadlock plants in Jamaica. Acalyphas are very popular in tropical landscapes as shrubs.

Cuttings root rather easily, though mist approximately every 30 minutes is helpful. High light, typically full sun or at most 47% shade, is required for good variegation and flowering. Acalyphas are very tender to cold, and temperatures should be kept above 50F (10C). Plants exposed to cold become very thin in appearance and usually need to be cut back to regain attractiveness.

Culture of copperleaf and chenille plants is very similar to that of *Hibiscus*. Both types require regular, consistent watering, or else older leaves turn yellow and drop. Two-spotted mite is the main problem encountered when growing these plants, especially on copperleaf. The density and angle of the foliage make spray coverage underneath the leaf difficult, so translaminar miticides, such as Avid (abamectin), are useful, as are the insecticidal soaps, if you can get good coverage. Root rot can occur rather quickly if drainage is inadequate.

AESCHYNANTHUS

Commonly known as the lipstick plant because of its red-orange flowers, *Aeschynanthus* has been popular in hanging baskets for many years. The two main varieties are *A. pulcher* and *A. radicans,* which has gained in popularity in recent years. The plants, with somewhat shiny, waxy foliage, have flowers that emerge from a purple tube (calyx). A yellow variety, *A. speciosum,* is also grown. Pot sizes range from 3 to 10 inches (7.5 to 25 cm), and hanging baskets range from 6 to 10 inches (15 to 25 cm).

Propagation is from cuttings taken from production baskets, which are rooted rather easily in the finishing container. Preferred light levels for good growth and flowering are 6,000 f.c. (64.6 klux), or about 55% shade. Soluble or liquid 3-1-2 ratio fertilizers are used, frequently at 200 ppm, with a leach every third or fourth watering. Optimum soil and air temperatures are from 70 to 80F (21.1 to 26.7C). Growth slows when soil temperatures dip below 65F (18.3C). Plants experience leaf drop and a reddening of foliage when the air temperature reaches 50F (10C). Because of their rather narrow temperature range, culture is best suited to greenhouses with adequate temperature control.

The most common disorder in the lipstick plant is *Corynespora* leaf spot, a small, blackish leaf spot that causes curling and distortion of the foliage. Keep leaves dry and spray occasionally with Daconil (chlorothalonil) to control it. *Botrytis* occurs occasionally on spent flowers or leaves, generally in cool weather. Clean up any plant debris, and spray with Chipco 26019 (iprodione). Maintaining a warmer greenhouse also helps. Iron deficiency appears as yellowing in the newer leaves, especially when roots are weak or soil aeration is poor. A drench with chelated iron generally solves the problem, in conjunction with careful watering. For interiors, bright indirect light of at least 500 to 800 f.c. (5.4 to 8.6 klux) is needed to maintain good form. Brighter light encourages flowering. Let the soil become fairly dry before irrigating.

The best definition of tropical is a place where coconut trees grow to maturity.

ARDISIA

The common name "coralberry" is applied to this plant because of the bright red berries, which are produced even when plants are rather small. Ardisias are common in smaller pots from 2 to 6 inches (5 to 15 cm), with the bulk of production produced in smaller sizes of 2 to 3 inches (7.5 cm). Small ardisias are also frequently seen in dish gardens.

Ardisias are grown from fresh seed, which is harvested during the fall and winter months. Germination rate should be at least 90% if seed is fresh and temperatures are maintained between 65 and 85F (18.3 and 29.4C).

Seedlings for small pots are frequently grown in straight sphagnum peat, whereas the larger sizes do well in a mixture of 60% peat, 30% pine bark, and 10% sand, by volume. Two- to 3-inch pots are grown at 1,500 to 2,000 f.c. (16.1 to 21.5 klux), or about 80% shade. Interestingly, larger specimens of 6 to 8 inches (15 to 20 cm) are grown brighter, usually at 73% shade, or 2,000 to 3,000 f.c. (21.5 to 32.3 klux). Under the bright light the plants tend to set fruit and have smaller leaves. Flowering usually occurs in June from seedlings planted the previous winter.

Disorders are relatively few, though seedlings develop curious stem galls under high temperatures. The heat kills a bacterium which the young coralberry plant needs in order to generate cytokinins. Without these hormones the young plants remain dwarfed and stunted, with galls on the stems.

ASPIDISTRA

Native to Japan, the cast-iron plant has long been a favorite of black-thumb gardeners everywhere. The plant tolerates a wide range of soils, temperatures, and light levels. Divisions of clumps are potted directly into containers ranging from 6 inches (15 cm) all the way up to 17-inch (42.5-cm) tubs, though 10-inch (25-cm) containers are probably the most popular size. Variegated forms are occasionally seen.

Many books say to grow *Aspidistra* somewhat brighter, but, at least in tropical conditions, you will do better under 80 to 90% shade, or less than 2,000 f.c. (21.5 klux). Cast-iron plants take more light under cooler temperatures. Maintain temperatures above 50F (10C), or browning and yellowing of the foliage will occur. Well-drained, peat-based mixes are common, as are slow-release, coated fertilizers with a 3-1-2 or similar ratio of $N-P_2O_5-K_2O$.

Cast-iron plants are somewhat fluoride sensitive, and they may develop slight tip burn when irrigation water or fertilizers contain significant amounts of fluoride. They like magnesium a great deal; lack of magnesium appears as broad, marginal yellowing in the older foliage. Drench or top-dress with magnesium sulfate if these symptoms develop, as they frequently do on older plants. Iron-deficiency symptoms occur if pH rises to near 7.0. Diseases are generally not a big problem, though aspidistras occasionally have problems with spider mites and scale insects.

Cast-iron plants do well indoors under a north window with light levels of 100 to 150 f.c. (1.1 to 1.6 klux). They tolerate as little as 50 f.c. (0.5 klux), one of the lowest minimum light requirements of any foliage plant. Keep the soil barely moist and irrigate conservatively, especially in winter. Feed with a teaspoon of soluble 20-20-20 and one-half teaspoon Epsom salts per gallon (3.8 l) about every three months. It is no problem for aspidistras to remain pot bound for extended periods.

BAMBUSA AND PHYLLOSTACHYS

Bamboo is somewhat underused as a foliage plant, though medium to large specimens do very well under high-light interior situations. The majority of ornamental bamboos are from the genera *Bambusa* and *Phyllostachys*. Two of the popular varieties include the feathery bamboo, *B. vulgaris,* whose canes are striped with bright yellow and green, and the black bamboo, *P. nigra*. Over 1,600 bamboo varieties exist, but importation of new varieties is limited because of the one-year quarantine restriction on imported bamboo. One variety I think we will see in the foliage trade in the coming years is *Schizostachyum brachy-cladum* (will somebody please give this a common name?).

Bamboo is generally propagated from division or rhizomes. Most varieties grow quickly and sucker well if given adequate moisture, light, and fertilizer. Weekly soluble 20-20-20 is popular, as are topdressings of slow-release, coated 13-13-13 or 14-14-14. Most take cold weather well, as many bamboos are not of tropical origin. Soil should remain moist but not soggy. Bamboos should absolutely not be permitted to dry out. Most varieties do best under bright light, around 6,000 f.c. (64.6 klux). Spider mites and rust diseases are occasional problems.

Indoors most bamboos need filtered light at a minimum of 500 f.c. (5.4 klux), though significantly higher light results in better growth and vigor. Cool drafts can be tolerated, but don't let bamboos get too dry. Old canes should be thinned periodically with very sharp pruning shears or a small saw. Iron deficiency can be problematic, especially in winter.

BEGONIA

Some begonias are produced as houseplants primarily for their attractive foliage, rather than for their flowering characteristics. Generally classified as *Begonia multi rex-cultorum,* these plants were derived from a series of crosses between *Begonia rex* and various rhizomatous begonias. The iron-cross begonia, *B. masoniana,* is an interesting plant with hairy leaves each sporting a large, chocolate brown cross. The beefsteak begonia, *B. multi erythrophylla,* has dark green leaves that are bright red underneath. Beefsteak takes less humidity than other begonias.

Rex begonias are best grown in a climate-controlled greenhouse maintaining high humidity and temperatures above 60F (16C). Light levels should be from 2,000 to 3,000 f.c. (21.6 to 32.4 klux). Soluble 20-20-20 is a very common fertilizer for begonias, at 150 to 200 ppm constant feed. Leach with clear water about every other week. Plant quality is improved if calcium nitrate is occasionally used in the fertilizer program, though it cannot be mixed with 20-20-20.

Numerous foliar diseases attack begonias, so it is best to keep foliage dry as much as possible. Drip irrigation is useful for larger pots, as are ebb-and-flow irrigation systems. Begonias are very susceptible to atmospheric fluoride, so be careful if you happen to be growing begonias near a phosphate plant. It is best to grow begonias a little drier in the winter.

Interior conditions should remain cool but not cold, between 60 and 75F (16 and 24C). Bright indirect or curtain-filtered light of at least 400 f.c. (4.3 klux) helps maintain good color, vigor, and leaf size. Keep rex begonias barely moist indoors but with good humidity. Overwater a rex begonia, and it's finished. Avoid fertilizing during the winter months.

TABLE 26 **Leaf analysis rating standards for *Begonia semperflorens-cultorum* hybrids**

Nutrient (%)	Very low	Low	Medium	High	Very high
Nitrogen	<3.50	3.50–3.99	4.00–6.00	6.01–7.50	>7.50
Sulfur	<0.20	0.20–0.29	0.30–0.75	0.76–1.25	>1.25
Phosphorus	<0.20	0.20–0.29	0.30–0.75	0.76–1.25	>1.25
Potassium	<2.00	1.00–2.49	2.50–6.00	6.01–7.50	>7.50
Magnesium	<0.25	0.25–0.29	0.30–0.75	0.76–1.25	>0.25
Calcium	<0.60	0.60–0.99	1.00–2.50	2.51–3.50	>0.50
Sodium			<0.21	0.21–0.50	>0.50
(ppm)					
Iron	<40	40–49	50–200	101–1000	>1000
Aluminum			<251	251–2000	>2000
Manganese	<30	30–49	50–200	201–1000	>1000
Boron	<15	15–19	20–75	76–125	>125
Copper	<4	4–7	8–100	101–500	>100
Zinc	<20	20–24	25–200	201–1000	>1000

Sources: Institute of Food and Agricultural Sciences, Apopka, Florida; Dr. Benjamin Wolf, Fort Lauderdale, Florida.
Notes: Common name: wax-leaf begonia. Sample of most recent fully mature leaves, no petioles.

CARYOTA

The fishtail palm, *Caryota mitis,* is well known as an interior foliage plant. The common name comes from the unusual shape of the leaflet, which strongly resembles a fishtail. It comes from the humid rain forests and swamp forests of Southeast Asia. Pot sizes range from 10 to 17 inches (25 to 42.5 cm) and larger.

Most growers produce fishtails under 63 to 73% shade, or about 3,000 to 4,500 f.c. (32.3 to 48.4 klux). Growing them under higher light early helps with suckering, while lower light at finishing produces a rich, deep green plant. Granular or coated, dry fertilizers of approximately a 3-1-2 ratio are used, and the fertilizer should have magnesium as well as trace elements. It is helpful to incorporate minor elements, with emphasis on iron, into the potting mix.

Helminthosporium leaf spot is a fairly common problem in caryotas, appearing as numerous roundish, brown spots all over the foliage. Sprays of Dithane (mancozeb) or Chipco 26019 (iprodione) should control it. A bacterial leaf spot disease caused by *Pseudomonas* looks rather similar. Sprays of Phyton 27 (picro cupric ammonium formate) may help, as does keeping the foliage dry. Spider mites are fairly frequently encountered during fishtail production, and most growers spray preventively for them. Iron chlorosis is frequently seen on older pots. A topdress with Hampshire iron or a similar granular iron product usually gives very good response, as do drenches with chelated iron.

For interiors, filtered sun or bright indirect lighting is preferred. Fishtails can become quite tall, so aggressive fertilization is discouraged. Iron chlorosis develops if soil is too moist or roots are struggling. Let the soil dry down a fair amount, then water thoroughly. Inspect for spider mites frequently, especially on new plants, which may harbor spider mite eggs.

FITTONIA

Fittonias are commonly called nerve plants because of the fine, white or reddish veins running throughout the leaves. *Fittonia verschaffeltii* has reddish leaves and veins, whereas *argyroneura,* the silver-nerve variety, has light green leaves with white veins. Fittonias originally came from the Andes. They are common in terrariums as well as medium-sized containers and hanging baskets.

Use a light, sphagnum peat-based mix. It helps to incorporate a low rate of trace elements into the medium. Cuttings root easily with occasional mist. Greenhouse temperatures should be kept above 60F (16C), with light levels between 1,000 and 2,500 f.c. (10.8 and 27 klux). Light rates of about 150 ppm nitrogen from a soluble or liquid 3-1-2 ratio fertilizer work well. Keep fertilization conservative on fittonias.

There is no substitute for light.

Rhizoctonia causes fungal root rot or blight, especially under warm, humid conditions. Drenches with Terraclor (PCNB) or Banrot (etridiazole plus thiophanate methyl) help. Southern blight occasionally attacks during hot weather. Discard affected plants and treat surrounding plants with Terraclor. Bacterial leaf spot caused by *Xanthomonas* may occur when foliage is moistened frequently. Keep foliage dry and spray with Dithane (mancozeb) mixed with a copper fungicide. Biden's mottle virus causes small, somewhat irregular, distorted leaves, especially on silver-nerve fittonia. The virus is transmitted by aphids from weed hosts.

Indoors, try to maintain at least 50% humidity and temperatures from 65 to 85F (18 to 29C). Fittonias don't like cold drafts, though I enjoy one once in a while. In low humidity interiors it helps to keep fittonias on a tray of moistened pebbles to generate humidity. A north window exposure with a minimum of 150 f.c. (1.6 klux) works well. They require a little grooming sometimes and fertilization every two to three months with a light rate of soluble 20-20-20.

HOMALOMENA

Homalomenas are rather new to the foliage trade. They are tropical aroids native to India and Burma. Their self-heading habit and heart-shaped leaves make them look something like a cross between *Philodendron selloum* and an anthurium. Emerald Gem, the most popular cultivar, holds its heart-shaped leaves on long petioles.

Culture of *Homalomena* is similar to that of *Philodendron,* in that it does well under low light and has high magnesium requirements. Most liners come from tissue culture and are potted into containers ranging from 4 to 10 inches (10 to 25 cm). Shade levels of 73 to 80% are common, resulting in light levels of 2,000 to 3,500 f.c. (21.5 to 37.7 klux). Use a well-drained mix that doesn't hold too much water, with a pH between 5 and 6. Keep temperatures between 60 and 90F (15.6 and 32.2C). Cold injury is observed around 40F (4.4C); leaves turn reddish and become water soaked, ultimately turning brown and dying.

Emerald Gem develops better root systems with a drench of a high-phosphate starter fertilizer, such as 10-52-10 or 8-30-6. Foliar sprays of Epsom salts at 2 pounds per 100 gallons (0.9 kg per 400 l) help plants maintain good color, especially under higher light levels. Root rot can be a problem with homalomenas grown in heavy mixes or kept too wet. Let the soil dry between waterings and make sure soluble salts are not high. Most growers prefer to use slow-release 18-6-8 or 19-6-12 on Emerald Gem, though liquid feed is also popular for smaller containers.

Mealybugs may be a problem from time to time, though sprays or drenches of Diazinon are helpful. *Pythium* is the major root rot disorder, so many growers like to drench with Subdue (metalaxyl), along with a high-phosphate starter fertilizer, at planting.

Indoors Emerald Gem does best when kept fairly warm, at least 70F (21.1C). Bright indirect light near a window is beneficial, though homalomenas do well under low light conditions if fertility is kept low. They tend to stretch when growth is induced under low light. Keep the plants on the dry side. If light levels are adequate, fertilize about every three months with a teaspoon of 20-20-20 and one-half teaspoon of Epsom salts per gallon (3.8 l).

PILEA

Several types of *Pilea* are in the commercial foliage trade, including the aluminum plant *(P. cadierei)*, the artillery plant *(P. microphylla)*, and creeping Charlie *(P. nummulariifolia)*. The aluminum plant comes from Vietnam, while the other two are native to the American tropics. Pileas are fairly easy to grow and are popular in 3- to 6-inch (7.5- to 15-cm) pots and smaller hanging baskets.

The plants root rather easily from cuttings, and usually mist is not needed. When planting *Pilea* don't compress the potting medium. The mix should have a pH between 5 and 6 and should consist of at least 40 to 50% sphagnum peat. Light levels should be from 1,000 to 2,000 f.c. (10.8 to 21.6 klux). Most varieties are rather cold sensitive, suffering injury around 55F (13C). Constant liquid feed with a 2-1-2 ratio of N-P₂O₅-K₂O is used, with an occasional leach. It is important to keep foliage dry when growing *Pilea,* and for most varieties overhead irrigation should be avoided. They do seem to like to have a little space between plants for good light exposure and good air movement.

A wide range of insects and diseases are capable of attacking pileas, but if cultural factors are managed well, pests and diseases won't give you many problems. *Xanthomonas* can be troublesome on aluminum plants. Sprays with copper or Phyton 27 (picro cupric ammonium formate) may help. If plants become too dry, they tend to lose their older foliage. If light levels are too high, the aluminum plant and creeping Charlie have rather dull color. Nutritional disorders in *Pilea* are not very common because they are fairly fast to finish.

Pileas do well for interior situations if temperatures stay on the warm side, usually 65 to 85F (18 to 29C). Avoid exposing the plants to temperatures below 55F (13C). Remember, these plants are native to very humid areas. A minimum of 40% humidity is required, or leaf burn and drought stress symptoms may develop. Soil moisture levels should be kept reasonable, but problems will develop if plants are kept too wet. Try to keep them near curtain-filtered sunlight, if possible. Minimum interior light levels range from 150 to 250 f.c. (1.6 to 2.7 klux).

PLECTRANTHUS

The Swedish ivy, *Plectranthus australis,* is actually native to Australia. It is a very popular houseplant in Sweden because it will tolerate temperatures down to 35F (1.7C), hence the name. A main attraction is the shiny foliage, and the white flowers are an added bonus.

Being in the mint family, *Plectranthus* roots easily and likes plenty of irrigation. Most growers get their cuttings from existing hanging basket production. Growers generally stick multiple cuttings directly into the finishing container. Mist is occasionally applied under warm or bright conditions. Rooting is fast and at a high percentage. Production light levels are generally 3,000 to 4,000 f.c. (32.4 to 43.2 klux). Best growth is achieved when temperatures remain above 55F (13C). Constant liquid feed of 20-10-20 is popular, at either 200 ppm nitrogen constant feed or 350 ppm about once a week. Some growers like to supplement with a small amount of coated, slow-release fertilizer, such as 13-13-13 or 14-14-14. Troubles are relatively few with Swedish ivy, as long as it is irrigated consistently. Mealybugs are an occasional problem, but sprays of Mavrik (tau-Fluvalinate) help control them.

Indoors Swedish ivy tolerates light levels ranging from 100 to 4,000 f.c. (1.1 to 43.2 klux); however, bright indirect light between 200 and 1,000 f.c. (2.2 and 10.8 klux) is preferred. It seems to be most comfortable with temperatures from 60 to 75F (15.6 to 23.9C), though it tolerates near-freezing temperatures. Swedish ivy thrives best when supplied with regular watering and good light levels. If allowed to become too dry, the plants become pale green and suffer from leaf drop. If light is too low for an extended period, the plants tend to lose their older foliage and become thin in appearance.

RADERMACHERA

The China doll, *Radermachera sinica,* is in fact a Chinese native. It has been in the United States trade only since the early 1980s. The most popular growing methods are to use two liners in an 8-inch (20-cm) pot or three liners in a 10-inch (25-cm) pot. Propagation is from seed, but most growers purchase liners.

The potting mix is critical for good China doll production. It is best to use a largely sphagnum-peat-based mix with about 30% bark, frequently with some perlite or polystyrene beads. China dolls do not do well in a cheap soil mix. The plants have rather weak roots, so it is important to drench with fungicides frequently, unless you have very good control of soil moisture levels. It helps to drench with Subdue (metalaxyl) and a thiophanate methyl fungicide at potting. Slow-release 18-6-12 or 18-6-8 fertilizers work well on China dolls. Some growers like to use triple 14 or triple 13 for the initial fertilizer application, switching to higher nitrogen later. Liquid feed can also be used, but the coated fertilizers work better on this plant, in my experience.

Radermachera is more cold tolerant than it looks, and it seems to tolerate heat well, also. Best growth seems to be at 73% shade, or 3,000 to 3,500 f.c. (32.3 to 37.7 klux). The biggest mistake most growers make is to let China dolls stretch, and then try to control the stretch with B-nine. The trick is to spray with B-nine before they stretch and to control the growth as it develops. For best quality it is very important to spray B-nine (daminozide) at 2,500 to 5,000 ppm. Generally, two applications are required, sometimes more in the summer.

China dolls are susceptible to cyclamen mites, which cause foliage distortion. Sprays with Avid (abamectin) work well. Mealybugs and scale insects can also be troublesome. Sprays with Dycarb (bendiocarb) help control them.

Radermachera is sensitive to root knot nematodes, and it is also very sensitive to ethylene. When soil is waterlogged, the ethylene generated causes very severe leaf drop. You therefore need a very good soil mix that is both well drained and well aerated, without excessive moisture-holding capacity. Sprays with silver

Fig. 44 *Radermachera* (China doll) liners, Apopka, Florida. (Courtesy of Marshall Horsman)

thiosulfate or potassium salts of indolebutyric acid are helpful, but they are generally too expensive for growers to use.

China dolls are not particularly good houseplants because they are quite sensitive to being too wet or too dry. Minimum interior light levels range from 250 to 300 f.c. (2.7 to 3.2 klux). Higher light levels are, of course, desirable. The plant likes a relatively humid interior environment. Absolutely do not let water collect in a saucer under the plant, or the potential ethylene generation will cause very severe leaf drop. The sap from China dolls can irritate some people.

SAINTPAULIA

African violets have been in the trade since 1926, when Baron von Saint Paul found the plants growing along a riverbank in what is now Tanzania and sent seeds back to Germany. They come in many colors, though the darker colors are somewhat easier to grow. Most production is in 4-inch (10-cm) pots, frequently with wick irrigation, capillary mats, or ebb-and-flow systems.

Light levels range from 1,000 to 1,500 f.c. (10.8 to 16.1 klux), though some varieties do well at 800 f.c. (8.6 klux). The darker green varieties do better at the high end of this light range, while the paler green varieties should be grown a little darker. It helps to keep consistent temperatures between 65 and 86F (18.3 and 29.4C). Many growers like to use mixtures of Canadian peat, vermiculite, perlite, and perhaps charcoal. The mix should be limed to a pH between 5.5 to 6.5. Soluble salts should be kept low, but some growers like to incorporate a little bit of coated, slow-release fertilizer into the mix. The plant can be grown from leaf cuttings, but most production is from tissue culture. It takes about nine months to finish an African violet in a 4-inch pot, sooner with a good-sized liner.

Overwatering will kill an African violet. *Phytophthora* causes rot and death if plants are overwatered. *Erwinia* may also attack under wet conditions.

Also, if the difference between the irrigation water temperature and the leaf temperature is greater than 15F (9C), whitish water spots will develop on the leaves. It is best to keep African violet foliage dry, period.

African violets are also rather sensitive to cyclamen mites, which cause distorted, somewhat clubby leaves. They are sensitive to Chipco 26019 (iprodione), so I don't recommend it for violets. *Botrytis* can occasionally attack spent flowers. Ornalin (vinclozolin) or Exotherm Termil (chlorothalonil) helps control it, as does Daconil (chlorothalonil). Powdery mildew is sometimes a problem, and Strike (triadimefon) is useful against this fungus.

African violets have been popular as flowering indoor plants since they were first introduced. A minimum of 40 to 50% relative humidity is required, though 60 to 80% humidity is better. Again, trays of pebbles or other means of artificially increasing humidity help in dry areas. Many growers fertilize with a teaspoon of 20-20-20 per gallon (3.8 l) about once a month, though there are many African violet fertilizers out there that are higher in phosphorus. Plants can bloom in as little as 200 f.c. (2.1 klux), though you will do better at 800 f.c. (8.6 klux). For flowering try to keep *Saintpaulia* near an east or north window or under fluorescent lighting. Don't overwater the plants, and remove old flowers.

SINNINGIA

A close relative of the African violet, the gloxinia *(Sinningia)* is native to Brazil. Its fleshy, velvety leaves and striking flowers make it one of the more beautiful houseplants when grown well. Production is usually in azalea pots, either 4 or 6 inches (10 or 15 cm). Do not try to grow it in translucent pots or in full gallon pots.

Don't pot gloxinias the day you receive the liners. Open the boxes, but be sure to wait about three days before they are transplanted to let them adjust to their new surroundings. Use a peat-and-perlite potting mix limed to a pH of 6.0. It helps to plant the liners deeply to avoid having floppy plants later.

Fertilize very gently at first. Use a high-nitrate formulation, such as soluble 15-16-17, alternating periodically with calcium nitrate. Don't give *Sinningia* too much phosphorus. A strap-leaf disorder caused by lack of boron is rather common, so make sure you have boron in the feed program.

Gloxinias like high relative humidity and temperatures of about 80F (26.7) in the daytime, 65 to 70 (18.3 to 21.1C) at night. The best light levels are 2,000 to 2,500 f.c. (21.5 to 26.9 klux). Artificial lighting is helpful when growing gloxinias in northern winter greenhouses.

They respond very well to supplemental carbon dioxide in the atmosphere and like gentle air movement from fans. Most growers spray with a 0.10% B-nine (daminozide) solution two to three weeks after potting.

Gloxinias do well indoors near an east- or north-facing window. They also prefer a high humidity environment and do well in the shower or a well-lit bathroom. They may continue to bloom with good light. Excessively low light results in weak, floppy growth. Avoid cold, drafty areas, and fertilize gently with 1 teaspoon of 20-20-20 per gallon (3.8 l) about once a month.

STRELITZIA

The white bird-of-paradise, *Strelitzia nicolai*, has become quite popular in recent years as a large, upright interior foliage specimen. Somewhat like a banana in appearance, the white bird is a close relative of the regular bird-of-paradise, *Strelitzia reginae*. Two seedlings are often started per 6-inch (15cm) container, then later stepped up into either a 10-inch (25-cm), 14-inch (35-cm), or 17-inch (42.5-cm) container.

Native to South Africa, the white bird-of-paradise is fairly tolerant of many soil types, but it loves calcium. Seeds are soaked in hot water before planting. Many growers like to put the two plants perpendicular to each other in the container, rather than parallel, for a fuller effect. They are rather slow growing, so the larger containers are fertilized either with granular 3-1-2 ratio fertilizers with trace elements or with coated, slow-release fertilizers of similar ratios.

An angular bacterial leaf spot caused by *Xanthomonas* is very common on younger birds of paradise. They largely grow out of the problem as they mature. While young, however, birds need to be sprayed with Dithane (mancozeb) plus a copper fungicide, or with Phyton 27 (picro cupric ammonium formate). Iron chlorosis is common on younger plants, as well, as indicated by a severe yellowing of the emerging foliage. Iron sprays help somewhat, but drenches or topdressings with chelated iron work faster and more completely to cure the deficiency. Watch your soluble salts carefully, as this plant is rather sensitive. For good, firm growth it helps to keep calcium nitrate or chelated calcium in the spray program on a regular basis. Flowers generally won't form until the white birds are quite mature, but the foliage is attractive enough that they don't need to be in flower to sell.

Give white bird-of-paradise as much light as you can in the interior environment. Put it in the brightest area you have. (We grow white bird-of-paradise as a landscape plant in full sun in the tropics.) Maintain steady soil moisture but don't overwater. If you have enough light to encourage growth, then fertilize gently with soluble fertilizer every four to eight weeks. In low light situations give white birds very little fertilizer.

B-nine is reactive with other chemicals, so it is best used with a dedicated B-nine sprayer.

TOLMIEA

The piggyback plant, *Tolmiea menziesii*, is native to western North America. While popular as a foliage plant, it requires cool conditions and certainly is not tropical; we definitely don't grow it in Florida.

Potting mixes are usually a mixture of sphagnum peat and bark. You need a light mix with good drainage and aeration. Light requirements are rather high, from 3,500 to 4,000 f.c. (37.8 to 43.2 klux). Light rates of soluble 3-1-2 ratio fertilizers are used, generally only once a week or so. *Tolmiea* does well at temperatures from 55 to 70F (13 to 21C) in the daytime, cooler at night.

Piggyback plants are susceptible to root rot diseases, which may occasionally require fungicide drenches. Mealybugs and whiteflies can also be troublesome. While I haven't tried it, I suspect the insecticide Marathon (imadacloprid) would be a useful long-term whitefly control.

Tolmiea makes a good porch plant in cool climates. The best daytime temperatures are similar to those in production, with nighttime temperatures from 40 to 50F (4 to 13C). It does well under fairly low light of 400 to 500 f.c. (4.3 to 5.4 klux), though it actually tolerates as little as 50 f.c. (0.5 klux). Very few interior plants will survive in such low light. Keep your piggyback plants cool and dry, as they hate to be overwatered. A teaspoon of 20-20-20 per gallon (3.8 l) every two months keeps them green and growing.

ZEBRINA

Formerly known as *Tradescantia zebrina*, the Wandering Jew is now commonly known as *Zebrina pendula*. *Zebrina* means "striped" foliage, while *pendula* refers to its hanging nature. Also called the inch plant, the Wandering Jew is native to Mexico and Guatemala. Many popular varieties exist, including the red to dark red *Z. purpusii* and Quadricolor, which has metallic green leaves striped with red, green, and white.

Zebrina is fairly tolerant of many things, but it needs relatively good light for best production. Try to grow it at 3,500 to 4,500 f.c. (37.8 to 48.6 klux), or about 63% shade. Growing temperatures should be kept above 50F (10C). Wandering Jews tolerate heat fairly well. Fertilize with 200 to 250 ppm of 24-8-16 on a constant-feed basis, leaching every couple of weeks. Keep the soil relatively moist. Avoid excessive drying, or loss of older foliage will result.

Inch plants have relatively few insect pests, though spider mites occasionally bother them. The normal arsenal of miticides should take care of the mites, but do your best to spray on the underside of the foliage. Since most production is for hanging baskets, this is not terribly difficult. *Cercospora* sometimes causes a fungal leaf spot, which can be controlled by keeping the foliage dry and spraying with a thiophanate methyl fungicide, such as Cleary's 3336 or Domain. These fungicides or Banrot (etridiazole plus thiophanate methyl) can also be used as a drench against *Rhizoctonia* root rot. Zebrina is somewhat fluoride sensitive, so use low-fluoride irrigation water and fertilizer sources. Two viruses are known to attack this plant, but they are not very common.

Wandering Jews in hanging baskets are popular and durable in interior environments. Keep temperatures between 65 and 85F (18 and 29C). Chilling injury can occur at 55F (13C). They tolerate relatively low light, down as low as 150 to 250 f.c. (1.6 to 2.7 klux), though more light is preferable. If light is low, variegated types tend to lose color contrast. Give *Zebrina* about one-half teaspoon 20-20-20 per gallon (3.8 l) about every two months. Tip burn, a fairly common problem, can be caused by excessive soluble salts, dry conditions, low humidity, or fluoride accumulation. Make sure the plants have adequate moisture and humidity, and tip-burn problems will be minimized.

REFERENCES

Adrian, B. April 1997. Telephone conversation with the author.

Blessington, T.M., and P.C. Collins. 1993. *Foliage Plants: Prolonging Quality,* 34-35, 45-46, 127-128, 161-165, 172-173, 197-200. Batavia, Ill.: Ball Publishing.

Chase, A.R., and L.S. Osborne. 1988. Effect of Dursban and *Pythium splendens* on Growth of Rex Begonia. *Foliage Digest* (September): 4-6

Chase, A.R., R.T. Poole, and L.S. Osborne. 1986. Lipstick Plant. *Foliage Digest* (September): 4-6.

Clarke, T. 1974. Welcome to the World of Ballet Violets. *GrowerTalks* (May): 1-9.

Conover, C.A., and R.T. Poole. 1988. *Production and Use of Ardisia crenata as a Potted Foliage Plant.* CFREC-Apopka Research Report RH-88-15. Apopka: Central Florida Research and Education Center.

Ellison, D. 1996. *Cultivated Plants of the World,* 29. Brisbane, Queensland, Australia: Flora Publications International.

Haehle, R. 1989. Acalypha godseffiana. *Florida Nurseryman* (November): 55-57.

McConnell, D.B., and R.W. Henley. 1986. Plant Profile for Interiorscapers: Swedish Ivy. *Foliage Digest* (December): 7.

Osborne, L.S., A.R. Chase, and C.A. Conover. 1986. African Violet. *Foliage Digest* (July): 1-4.

Osborne, L.S., A.R. Chase, and R.W. Henley. 1986. Pilea. *Foliage Digest* (November): 3-6.

Peterson, S.S. 1969. *Success in Growing African Violets,* 1-7. Cincinnati: Sylvia Peterson.

Ridings, W.H., S.F. Fazli, and J.W. Miller. 1975. Temperature and Other Factors Affecting the Frequency of Galling in Ardisia Seedlings. *Proc. Fl. State Hort. Soc.* 88: 578-582.

Standard and Super Compact Gloxinias. 1981. West Chicago, Ill.: Ball Seed Co.

Stewart, L. 1994. *A Guide to Palms and Cycads of the World,* 72-73. Sydney, Australia: Angus & Robertson.

Sweet, J.S. 1979. *Gloxinias, 365 Days of the Year.* Ohio Florist's Assoc. Bulletin No. 596. Columbus: Ohio Florist's Association.

Thomas, S.H. 1992. Grower's Notebook: *Homalomena. Greenhouse Manager* (May): 12.

THE SECRET
TO BEING A GOOD FOLIAGE GROWER

ATTENTION TO DETAIL!

Fig. 45 Look closely. Pollinator at work in Puerto Rico. (Photo by Bryan Perrigo)

Index